Economic Instruments for Environmental Management
A Worldwide Compendium of Case Studies

Edited by
Jennifer Rietbergen-McCracken and Hussein Abaza

UNEP

Earthscan Publications Ltd, London

First published in the UK in 2000 by Earthscan Publications Limited
for and on behalf of the United Nations Environment Programme

Copyright © United Nations Environment Programme, 2000

All rights reserved

A catalogue record for this book is available from the British Library

ISBN: 1 85383 690 7

Typesetting and page design by PCS Mapping & DTP, Newcastle upon Tyne
Printed and bound by Biddles Ltd, Guildford and King's Lynn
Cover design by Susanne Harris

For a full list of publications please contact:

Earthscan Publications Limited
120 Pentonville Road
London N1 9JN
Tel: +44 (0)20 7278 0433
Fax: +44 (0)20 7278 1142
Email: earthinfo@earthscan.co.uk
http://www.earthscan.co.uk

Earthscan is an editorially independent subsidiary of Kogan Page Limited and publishes
in association with WWF-UK and the International Institute for Environment and
Development

This book is printed on elemental chlorine-free paper from sustainably managed forests

CONTENTS

EASTERN AND CENTRAL EUROPE REGION CASE STUDIES

LATIN AMERICA REGION CASE STUDIES

LIST OF TABLES

PREFACE

This compendium represents a three-year collaborative effort that has involved numerous research institutions from Africa, Asia, Eastern and Central Europe, and Latin America. In supporting and coordinating this work, UNEP has sought to provide an opportunity for researchers from developing countries and countries in transition (CITs) to review the use of economic instruments in their own regions, and learn from the shared experiences.

One of the most striking features of this collection of case studies is the diversity of environmental problems which have been targeted by these policy instruments, including air and water pollution, packaging, deforestation, overgrazing, and wildlife management. Clearly the scope for using market-based instruments for environmental purposes in these countries is very great – perhaps even more so than in developed countries.

And yet, the effectiveness of these economic instruments has been limited, as the same shortcomings seen in many developed countries' applications have been repeated by developing countries and CITs, as evidenced in this book. Most notable among these limitations is the failure to set environmental charges at a level high enough to induce behavioural change among polluters or users of natural resources. This means that while the charges may serve a useful revenue-raising function, their potential incentive impact is greatly restricted.

This compendium therefore provides a key source of empirical evidence to show not only the potential power of economic instruments for environmental management, but also the main pitfalls to avoid in introducing these instruments. By learning from the kinds of pioneering experiences documented here, the future use of economic instruments can be even more successful.

ACKNOWLEDGEMENTS

This long-term project involved a great many people who deserve to be thanked, although space does not allow them all to be mentioned here.

The editors would like to thank the regional coordinators for their participation in this project and their willingness to assist in finalizing the chapters for publication – in particular, N H Fidzani (Africa); Adis Israngkura (Asia); Zsuzsa Lehoczki (Eastern and Central Europe); and Ronaldo Seroa da Motta (Latin America). In UNEP, Deborah Vorhies and Ivonne Higuero were responsible for the overall coordination of this project and Naomi Poulton undertook the initial editing work.

The editors would also like to thank Frank Vorhies (Head of Environmental Services Unit, IUCN, Gland) and Ali Dehlavi (Economics, Trade and Environment Unit, UNEP) for their kind assistance, and Simon Rietbergen for his support and advice, and wish to publicly absolve them of any responsibility for any remaining errors.

INTRODUCTION

Jennifer Rietbergen-McCracken

BACKGROUND

This compendium is the product of several years' collaborative work by researchers in a wide range of developing countries and countries in transition (CITs), facilitated and coordinated by the Economics, Trade and Environment Unit of the United Nations Environment Programme (UNEP). Collaboration began in 1995, when UNEP convened an expert group meeting to discuss how best to support the use of economic instruments and valuation methods in these countries. It became clear that, while economic instruments were being increasingly used in many developing countries and CITs, there had been little documentation of these experiences and hence little opportunity to share the lessons that were being learned. At the same time, it was becoming evident that the use of market-based instruments in OECD countries was often not directly transferable to developing countries and CITs, where prevailing economic, social, and political conditions strongly influence the way in which these instruments can be used. Furthermore, many of the environmental problems facing these countries are quite different from those targeted by economic instruments in developed countries.

It was therefore decided to produce a set of case studies showing the possibilities and practicalities associated with using economic instruments in developing countries and CITs. Research institutions were chosen from each of the four regions – Africa, Asia, Eastern and Central Europe, and Latin America – to help coordinate the selection and production of the case studies. Several rounds of discussion meetings and workshops brought together the regional coordinators to develop a consistent approach and ensure that the cases covered a range of different applications.[1] A common framework for the analysis was provided by Panayotou (1994). Most of the chapters were drafted between 1995 and 1996 and reviewed during 1996 and 1997. Because of the inevitable delays in finalizing the output of such a collaborative effort, and the rapid pace of change in the situations described in the cases, a final round of updating was done in early 1998. A companion compendium on the use of *environmental valuation methods* in developing countries and CITs is also available (Rietbergen-McCracken and Abaza, 2000).

OBJECTIVE AND AUDIENCE

The compendium is intended to serve two main purposes, one connected with the 'demand' side, the other with 'supply'. On the demand side, the book will help fill the considerable gap in the existing literature on economic instruments, with respect to the application of these instruments in developing countries and CITs. Up until now, policy makers and economists in these countries have had to rely primarily on (often inappropriate) examples from the developed world for guidance. On the supply side, the book is intended to provide a forum for developing country and CIT researchers to air their views and perspectives. The field of environmental economics as a whole is still largely dominated by western academics, and local researchers have had little opportunity to reach an international audience. By involving national research institutions within the four regions, UNEP has purposely tried to correct this imbalance, while at the same time encouraging networking between the different institutions.

The main audience for this book comprises policy makers and economists in developing countries and CITs, who are looking for practical descriptions of and guidance on how economic instruments can be applied, and what their likely impacts will be. The cases will also provide useful material for researchers and university lecturers in the fields of environmental economics and environmental policy. Readers without formal training in economics should not feel unwelcome. The book has also been written with them in mind, and this introductory chapter and the extensive glossary should help them make their way through (or round) the economics involved.

SCOPE

The range of economic instruments available to policy makers is in fact quite broad. Panayotou defines several types of economic instruments for environmental management, including:[2]

1   *Redefining Property Rights*
    • changes in ownership, use and development rights.
2   *Market Creation*
    • tradeable emission permits.
3   *Liability*
    • liability insurance legislation.
4   *Charge Systems*
    • effluent charges;
    • user charges;
    • product charges;
    • administrative charges;
    • impact fees;
    • access fees.
5   *Fiscal Instruments*
    • pollution taxes;
    • input taxes;
    • import tariffs;
    • financial aid in installing new technology;
    • subsidies for environmental research and development expenditure.
6   *Deposit-Refund Systems and Bonds*
    • deposit-refund schemes to encourage recycling;
    • environmental performance bonds;
    • land reclamation bonds.
7   *Financial Instruments*
    • financial subsidies
    • soft loans and grants;
    • sectoral/revolving funds.

In compiling the 19 different case studies for this compendium, effort has been taken to try and cover the different types of economic instrument currently in use in the regions concerned. Table I.1 lists the instruments described in the cases and how they fit in the above classification. Clearly, the majority of cases deal with some form of environmental tax or charge – the most common type of economic instrument in use in these regions (see Achanta et al, 1995).

Table I.1 *Range of Economic Instruments Covered*

Country	Environmental Objective	Category of Economic Instrument		
		Redefining Property Rights/Market Creation	Charge Systems	Fiscal Instruments
Africa				
Botswana	Rangeland management	Change in land tenure		
Botswana	Management of forest reserves		Stumpage fees	
Botswana	Management of water consumption		Water user charge	Subsidies for rural users
Zimbabwe	Wildlife conservation	Decentralization of wildlife user rights		
Ghana, Botswana, Zimbabwe	Human and solid waste management in urban areas		User charge for waste disposal and treatment	
Asia				
Malaysia	Reduction of palm-oil-related water pollution			Pollution tax for palm-oil processing mills; subsidies to environmental R&D
Thailand	Reduction of lead in the environment			Unleaded gasoline price differential
Thailand	Reduction of groundwater extraction		Groundwater user charge	
Thailand	Wastewater management in industrial parks		User charge for waste-water treatment	
Thailand	Maintenance and protection of a national park		Entrance fees	

Country	Environmental Objective	Category of Economic Instrument		
		Redefining Property Rights/Market Creation	Charge Systems	Fiscal Instruments
Eastern and Central Europe				
Poland	Reduction of air pollution		Sulphur dioxide emission charge	
Hungary	Solid waste management		Product charge on packaging	
Hungary	Reduction of air pollution		Product charge on transport fuel	
Poland	Reduction of water pollution		Water pollution charge	
Hungary	Management of water resources		Water abstraction fee	
Latin America				
Mexico	Reduction of water pollution			Wastewater effluent charge
Colombia	Reduction of water pollution		Water pollution tax	
Chile	Reduction of air pollution in Santiago	Tradeable air quality permits		
Brazil	Reforestation (national)			Reforestation tax
Brazil	Forest conservation (Minas Gerais)			Forestry tax
Brazil	Nature conservation (3 states)			Fiscal compensation to municipalities with land-use restrictions

ECONOMIC INSTRUMENTS VERSUS REGULATIONS?

The popularity of environmental taxes and charges with governments in developing countries and CITs is largely due to these instruments' potential for generating win-win situations. That is to say, the charges imposed can benefit the *treasury* (by raising revenue in a cost-efficient manner), the *environment* (by encouraging polluters or users of environmental resources to change their behaviour to become less polluting or wasteful, and by using the revenue raised to support environmental protection efforts), and the *economy* (by creating incentives for private sector investment in the development of cost-effective environmental protection technologies). This advantage of economic instruments contrasts with the more traditional command-and-control approach, based on regulations and standards, which can be administratively costly to implement, with no real incentive function and sometimes with no direct revenue-raising function either.

The two other categories of economic instrument covered in this compendium – redefining property rights and subsidies – also offer significant advantages over regulations. The former provides incentives, in the form of more secure ownership or user rights, for polluters or environmental users to control pollution or manage the resource sustainably; the latter helps to bridge the gap between the private cost and social cost of pollution control or sustainable resource use.

However, this does not mean that economic instruments can or should replace regulations. Kahn (1995) identifies three sets of circumstances where command-and-control policies may be more desirable than economic instruments:

1   when monitoring costs are high;
2   when the optimal level of emissions is at or near zero; and
3   during random events or emergencies that change the relationship between emissions and damages.[3]

Several cases provide examples of how both forms of policy instrument can operate well in combination with each other. For example, in Chapter 8, Israngkura shows how the conservation of groundwater in Thailand, through the introduction of a user charge, has been strengthened by the imposition of a total ban on groundwater extraction in certain critical areas. Similarly, in Chapter 4, Arntzen describes how in Zimbabwe the decentralization of wildlife user rights to local communities has been done in conjunction with the establishment of conservative hunting quotas, to avoid over-exploitation of the wildlife resources. In other cases, the market-based instruments have been introduced as a temporary measure, to be replaced later by regulations. This is the case with the packaging product charge in Hungary, described by Balogh and Lehoczki in Chapter 12. There are plans to replace this charge with a set of legal obligations for recycling and treatment of packaging, once the necessary recycling infrastructure has been built, with the financial support of the environmental fund to which the charge revenue is being channelled.

It has sometimes proven difficult to get the right balance between market-based and regulation-based policy instruments. The environmental impact of the sulphur dioxide charge in Poland, for example, has been greatly limited by the very strict emission limit and hefty fines imposed by the associated permit system (Chapter 11 by Lehoczki and Sleszynski). These two regulations effectively mean that polluting firms have no legal choice, and very few technological options, for exceeding their emission limit. And, because of the considerable problems encountered in the design and implementation of economic instruments (as discussed below), it has sometimes been left to the command-and-control policies to effect the desired environmental improvement. In Santiago, Chile, improvements in air quality have come, not so much from the tradeable emission permit system, but from the air quality standards required of the firms before they can participate in any permit transactions (Chapter 18 by Seroa da Motta and Rios Behrem).

APPLYING ECONOMIC INSTRUMENTS IN DEVELOPING COUNTRIES AND COUNTRIES IN TRANSITION

The case studies in this volume provide a wealth of knowledge about the practical issues involved in the application of economic instruments in these countries. Again and again, they present evidence of the need to address the particular situations of developing countries and countries in transition, where conditions are often quite different from those found in more developed countries. In comparison with most developed countries, many developing countries and CITs face:

- more severe environmental degradation;
- a greater reliance on environmental resources for their economic development;
- weak institutional capacity to implement economic policy;
- greater risks of public protests and political resistance to the introduction of environmental charges;
- important issues of equity and social justice, particularly in countries with highly skewed income distributions; and
- weaker environmental research and development capacity.

At the same time, the potential benefits of economic instruments to these countries make it even more important to recognize and address these issues, to help ensure these instruments are used appropriately and effectively. The major issues involved can be categorized into three groups: design-related, implementation-related, and impact-related issues. It should be borne in mind that the application of economic instruments in developed countries has also encountered problems in these three areas.[4]

DESIGN-RELATED ISSUES

It is in the preparation and design of economic instruments that *political pressures* have had their greatest impact. Since many of the instruments described in the cases are the first such market-based instruments to be introduced in the countries concerned, their preparation has often involved lengthy and heated political negotiations. In the case of environmental charges, these negotiations have often resulted in a weakening of the charge, as governments bow to pressures from industry to reduce the charge rates or introduce exemptions for particular categories of users or polluters. This was evident, for example, during the preparation of the product charge on transport fuel in Hungary (Chapter 13 by Lehoczki), and in this case the political discussions resulted not only in a lower charge rate, but also in the removal of the proposed differentials for leaded and unleaded gasoline. This was judged necessary, since the even lower rate for unleaded gasoline would have made it insignificant.

Despite these kinds of problems associated with negotiating the format of economic instruments with those groups likely to be affected by them, there are clearly some advantages in doing so. By involving a wide range of stakeholder groups (including, as appropriate, industry, NGOs, user groups and consumer associations), policy makers can learn about what is politically acceptable, and can begin the necessary process of information-sharing and consensus-building with those whose cooperation will be crucial to the success of the instrument's application. In Chapter 17 on the water pollution charge in Colombia, Seroa da Motta, Rudas and Ramírez even claim that the participation of water users may also replace the need for an economic valuation of environmental resources and costs. They propose that an agreement over environmental goals may reflect the value a community places on damages from pollution.

While this view may not be shared by all the other authors, it is clear that stakeholder participation in decision-making has played an important role in several of the cases. One example comes from the product charge on packaging in Hungary (Chapter 12), where a technical committee was established to help formulate the product charge, propose mechanisms for its implementation, and decide how the revenue should be spent. This committee, which meets at least every three months, includes seven delegates from professional associations, two from environmental NGOs, two from local government, one each from seven different ministries, and a representative of the Consumer Protection Agency.

In addition to setting charge rates according to what is deemed *politically* acceptable, policy makers have also based rates on particular *environmental* or *economic* objectives. For instance, the level of the palm oil pollution tax in Malaysia was set based on the objective of reducing the average effluent discharges from palm oil mills to about one-tenth of the original discharge levels (Chapter 6 by Israngkura). And, although not officially acknowledged, the level of the water pollution charge in Poland is set based on the revenue requirements of the water treatment plants (Chapter 14 by Lehoczki and Sleszynski).

Setting environmental charge levels in this way, ie according to political, environmental or economic bottom lines, obviously precludes the possibility of the charges achieving *economic efficiency*. To be economically efficient, an environmental charge would need to equate the marginal social cost of reducing pollution by an additional unit with the marginal benefit of one unit reduction of pollution (ie the value of averted damages). In other words, the charge level would need to be set such that it would induce polluters to implement abatement measures up to the point where the marginal benefit from pollution reduction equals the marginal cost of doing so. None of the environmental charges described in this compendium appears to fulfill this criterion, which is hardly surprising, not only because of the environmental, political and economic factors mentioned above, but also because of the considerable difficulties involved in determining the economically efficient level.[5] Probably the only exception is the retributive water pollution tax in Colombia, which has been established as a pigouvian (ie economically efficient) tax in the relevant legislation, although its present application is based on pre-determined environmental targets. The long-term aim is to apply valuation techniques to put a value on environmental damages, to enable the tax rates to get closer to the optimum levels defined by law. However, there are considerable doubts as to the feasibility of achieving this (von Ameberg, 1995, cited by Seroa da Motta et al in Chapter 17). It should be noted here that there are very few, if any, cases of economically efficient environmental charges being applied in developed countries (Tietenberg, 1992).

IMPLEMENTATION-RELATED ISSUES

The most obvious source of implementation problems is the weak institutional capacity within many developing countries and CITs, which limits the effectiveness of the economic instruments involved. The actual institutions responsible for administering the instruments described in this compendium include government ministries and departments (including those dealing with the environment, mineral resources, forestry, wildlife, and health); Environmental Protection Agencies at the national or provincial levels; central and regional public water bodies; tax collection offices and customs offices. In several instances, a new institution has been set up to oversee the collection of the environmental charges and the disbursement of the revenue collected. These institutions typically suffer from a lack of resources and experienced personnel to enable them to undertake their responsibilities, and the usual result is a weak enforcement and monitoring of charge obligations, and hence a much lower revenue level than that which was anticipated. To name but one example, the revenue collected from the wastewater effluent charge in Mexico has been greatly limited by the weak enforcement capacity of the National Water Commission (the Spanish acronym of which is CNA), the body responsible for applying the charge. In fact, the charge revenue collected by CNA in 1994 represented only 2 per cent of the potential revenue collection level (Chapter 16 by Seroa da Motta, Contreras and Saade).

Many of the countries applying environmental charges rely on self-reporting of the consumption or pollution levels by the users or polluters themselves. This is a necessary mechanism, given the overwhelming monitoring load that would otherwise be placed on the institution administering the charge. However the predictable result is the commonplace under-reporting of consumption and pollution levels, and the subsequent reduction in the amount of revenue collected. Some institutions do manage to undertake random monitoring visits to check the reported levels against the actual levels, but staff shortages make these controls very patchy. One strategy, as adopted by the beleagured CNA in Mexico, is to be more selective in the monitoring efforts. CNA is now focusing inspections largely on industrial dischargers and water users in cities of more than 10,000 inhabitants, where the highest concentrations of discharges are found.

While most of the cases in this compendium focus their analysis of implementation issues on the performance of the relevant public institutions, the capacity of the private sector also plays a role, particularly in the development of environmental technologies.

IMPACT-RELATED ISSUES

In assessing the environmental impacts of an economic instrument, one needs to consider both its *incentive* impact (ie the extent to which it has induced behavioural change among polluters or users of environmental resources) and its *revenue-spending* impact (ie the extent to which the charge revenue has supported environmental protection or clean-up efforts). By far the most common finding in the case studies, as far as the instruments' incentive impacts are concerned, is that the charge rates were set too low to effect any significant change. This is particularly evident in the case of the national reforestation tax in Brazil, which was introduced as an alternative option for timber users, who could either pay the tax or, as they had previously been required to do, undertake reforestation activities themselves. The fact that all those who are eligible to pay the tax have opted to pay it rather than invest in reforestation suggests that the tax rate was too low in comparison to the real costs of reforestation (Chapter 19 by Seroa da Motta).[6]

Where reductions in pollution or consumption levels have been seen, it is extremely difficult to distinguish the impact of the charge from the impacts of other policy measures (such as legislation on emission limits, or the removal of subsidies) or economic changes which have occurred since the charge was introduced. Take for example the water pollution charge in Poland described in Chapter 14. There has clearly been a significant decrease in the amount of wastewater discharged (from nearly 13 billion m^3 in 1985 to less than 10 billion m^3 in 1994). However, this coincided not only with the introduction of the pollution charge but also with the introduction of a wastewater discharge permit system and a period of economic decline following transition – both of which are likely to have contributed to the observed environmental improvement.

In a few cases, however, it is possible to directly attribute a change in environmental behaviour to an economic instrument. In the state of Minas Gerais in Brazil, the introduction of a forestry tax has obviously been a key factor in changing the pattern of charcoal consumption in the state (Chapter 19). This is due in large part to the fact that the tax is differentiated according to species and uses of timber, with the use of native forest timber for charcoal production being particularly penalized. Unfortunately, the reduction in logging of native forests in Minas Gerais has been accompanied by an increased exploitation of forests in neighbouring states where such heavy taxes are not being applied.

The environmental impact from the *use of revenue* is easier to measure and has generally been successful. The revenue of many of the environmental charges is channelled directly into environmental funds at the national, regional or municipal levels, or is transferred to public institutions responsible for pollution control. As Chapter 19 shows, a particularly interesting form of economic instrument is being experimented with in Brazil, involving fiscal compensation for municipalities that maintain land-use restrictions for environmental purposes. Here, some states are setting aside a certain percentage of value added tax revenue to compensate those municipalities where land-use restrictions are in place, and to encourage other municipalities to implement similar sustainable activities. The amount and destination of the revenue from this instrument are easily tracked, and it was estimated that in 1994 some US\$127 million were generated in the three participating states.

The major problem encountered in spending the revenue has been an initial slow rate of disbursement due to institutional problems or, in a few cases, excessive earmarking to specific projects.[7] In one case – the national reforestation tax in Brazil – the central government agency responsible for spending the revenue actually absorbed all the funds itself, to deal with its own financial crisis, rather than spending the money on reforestation activities, as was intended. In Chapter 19, Seroa da Motta explains how this situation has only recently been addressed by allowing part of the charge revenue to be diverted to states and NGOs willing to invest in reforestation.

Earmarking seems to present something of a dilemma, as some authors see it as a necessary means of guaranteeing financial support for efforts to address the environmental problem targeted by the charge, while others report that excessive earmarking has decreased the efficiency of the charge and slowed down disbursement.

In comparison to their *environmental* impacts, the *economic* impacts of economic instruments are, on the whole, more obvious and easier to measure. The vast majority of the charges described in this book have been more successful in raising revenue than in inducing behavioural change, and in some cases revenue-raising has been explicitly recognized as the priority objective of the charge. It should be remembered that the majority of economic instrument applications in developed countries have also focused on revenue-raising rather than behavioural change (Panayotou, 1998). Furthermore, some economists (eg Pearce and Warford, 1993) argue that the revenue-raising function of environmental charges in developing countries is often the only feasible one, as incentive impacts would generally require politically unacceptably high charge rates. Here again though, the low rate of most of the charges, coupled with the frequently reported problems of enforcement and non-compliance, has limited their revenue-raising performance.

Looking at the *distributional* impacts, some of the cases have highlighted important issues of *equitability*. Two cases – the gasoline price differential in Thailand described by Israngkura in Chapter 7, and the change in land tenure in Botswana described by Fidzani in Chapter 1 – actually show that the instruments had a negative effect on equitability, by favouring the richer segment of society and placing a heavier burden on the poor. The lower price for unleaded gasoline favoured the owners of newer cars with fuel injection systems that require the use of unleaded gasoline. Lower income owners with older cars, equipped with the old carburetor system that is unsuitable for unleaded gasoline, had to continue paying the higher price for leaded gasoline. The equitability problem in Botswana was considerably worse. Here, the change in land tenure was intended to discourage overstocking and overgrazing, to protect the rangeland and stop large-scale herders from crowding out those with small herds. In reality, instead of reducing the gap between rich and poor livestock farmers, the policy actually widened it, by transforming large areas of previously communal rangeland into private ranches for large-scale farmers, and charging extremely low rents. At the same time, the policy allowed these farmers to have continued access to the remaining commons, to compete for grazing with the small-scale herders. The setting aside of land for the private ranches also displaced groups of hunters and gatherers whose livelihoods depended on access to the land.[8]

In some instances, environmental charges have been designed to ensure they do not adversely affect the poor. Again in Botswana, the water user charge rate was differentiated between rural and urban users and between low and high consumption levels, with additional subsidies being provided to rural water users (Chapter 3 by Arntzen). This policy is viewed as successful in promoting social justice, although the author recommends that the more wealthy rural users pay the same as urban users and that the subsidies be better targeted to improve the efficiency of the charge while still protecting the poor.

A final word on the impacts of economic instruments. Most of the instruments described in this compendium have been in operation, in their present form, for less than eight years; some were introduced as recently as two or three years ago. It is therefore too early to judge what impacts they will likely achieve in the medium to long term. Many of the instruments will no doubt be revised, expanded or discontinued, depending on their performance, and the institutional, environmental and economic conditions will continue to change too.

THE PARTICULAR CASE OF COUNTRIES IN TRANSITION

While many of the issues mentioned so far apply equally to developing countries and CITs, it is worth looking at a few particular characteristics of CITs which have affected how economic instruments have been introduced and applied. It is evident from the cases from Eastern and Central Europe that the transition to a market economy has not only made possible the introduction of market-based instruments, but has also presented a number of challenges for their successful implementation.[9] The packaging charge in Hungary (Chapter 12) is a good case in point. Here the environmental problem – the increasing use of packaging material and the resultant problems of its disposal – was actually a result of the country's transition, as marketing took off and packaging became an important marketing tool. Transition also resulted in the loss of the legal basis for the deposit-refund system that had previously been in place, hence the need to devise an environmental charge to tackle the environmental damage from packaging. Still in Hungary, it was only after transition that the severe problems of overexploitation of the country's water resources were revealed, and a user charge was introduced to try and better reflect the increasing value of this scarce natural resource (Chapter 15 by Burger and Lehoczki).

In Poland, on the other hand, the environmental charge system had already been introduced under the conditions of a centrally-planned economy, so in this case the transition process did not create or reveal the need for the charge, but rather made the costs of the charge to industry *real*. Previously, the state-owned enterprises could rely on the state to cover the additional costs they incurred from these charges. This had meant that there was little opposition from these enterprises to the introduction of the charges. The transition, however, passed these costs back to the companies themselves, who reacted with strong protests. In 1992 a new incoming government responded to the protests by drastically reducing all emission charge rates. This move also provoked strong protests, from industries who had already started abatement measures and the public who saw a decline in the revenue available for the environmental funds. The following year the government reestablished the rates at the previous levels.

As already mentioned, the recession period accounted for substantial improvements in air and water quality, as the output from polluting industries declined. The harsh economic conditions also made it harder for companies to comply with the new environmental charges. This has been particularly so for the older companies in heavy industry (such as coal and steel), which had previously survived on state subsidies that were removed during transition. This ongoing problem of non-payment has caused considerable difficulties in realizing the economic and environmental objectives of the charges.

OUTLOOK FOR THE FUTURE

The problems discussed in this chapter, relating to institutional weakness, economic and political pressures, and other shortcomings in how the instruments have been applied, are to be expected, especially considering the ground-breaking nature of these first applications. Many of the instruments were introduced during or just after major economic and political upheavals, and the necessary institutional capacity and administrative framework had to be built, often from scratch. The future implementation of these charges, and the introduction of new ones, should become considerably easier, as the institutional foundations are already in place, the political will has been mobilized, the economic rationale has been justified, and the environmental awareness (and public relations concerns) of industries is increasing. And, as the experiences of these countries (including many not covered in this book) are shared, the use of economic instruments is likely to grow rapidly in the next few decades.

NOTES

1   Contact information for the regional coordinators is provided at the end of this chapter.

2   Adapted from Panayotou, 1998.

3   That is, events that temporarily change the pollution damage function. For example, if a drought reduces the volume of water in a river, then each unit of pollution that is discharged into the river is less diluted and causes greater damage (Kahn, 1995).

4   For a recent review of the performance of economic instruments in developed countries, see OECD, 1997

5   These difficulties include estimating the incremental cost of reducing pollution by alternative means, and valuing marginal environmental improvements. Another potential difficulty stems from uncertainty about the shape of the actual pollution damage function. See Kahn (1995) for a discussion of different shaped marginal damage functions.

6   This situation may however still be better than the prior situation of almost total non-compliance with command-and-control regulations regarding reforestation.

7   See glossary for a brief description of earmarking, as applied here.

8   The privatization of rangeland in Botswana also had disastrous effects on the wildlife populations, huge numbers of which died after their migration routes were blocked by the fenced-off ranches.

9   See Pearce and Warford (1993) for an analysis of the main challenges involved in applying economic instruments in Eastern Europe.

REFERENCES

Achanta, A N, Mittal, M, and Mathur, R (1995) The Use of Economic Instruments in Carbon Dioxide Mitigation: A Developing Country Perspective. *Environment and Trade Series* No 12, United Nations Environment Programme, Geneva.

Kahn, J R (1995) *The Economic Approach to Environmental and Natural Resources*. The Dryden Press, Harcourt Brace College Publishers, Orlando.

OECD (1989) *Economic Instruments for Environmental Protection*. OECD, Paris.

OECD (1997) *Evaluating Economic Instruments for Environmental Policy*. OECD, Paris.

Panayotou, T (1994) Economic Instruments for Environmental Management and Sustainable Development. Report submitted to United Nations Environment Programme. Harvard Institute for International Development, Massachusetts.

Panayotou, T (1998) *Instruments of Change: Motivating and Financing Sustainable Development*. Earthscan Publications, London.

Pearce, D W and Warford, J J (1993) *World Without End: Economics, Environment, and Sustainable Development*. The World Bank, Washington, DC, and Oxford University Press, New York.

Rietbergen-McCracken, J A and Abaza, H (eds) (2000) *Environmental Valuation: A Worldwide Compendium of Case Studies*. Earthscan Publications, London.

Tietenberg, T H, (1992) Economic Instruments for Environmental Regulation In Markandya, A and Richardson, J (eds) *The Earthscan Reader in Environmental Economics*. Earthscan Publications, London.

von Ameberg, J (1995) Selected Experiences with the Use of Economic Instruments for Pollution Control in Non-OECD Countries. In Borregar, N et al (eds) *Uso de Instrumentos Econômicas en la Politíca Ambiental: Análisis de Casos para una Gestión Eficiente de la Contaminación en Chile.* CONAMA, Santiago.

CONTACT INFORMATION FOR REGIONAL COORDINATORS

N H Fidzani (Africa Region Coordinator)
University of Botswana
Private Bag 0022
Gaborone
Botswana
email: fidzannh@noka.ub.bw

Adis Israngkura (Asia Region Coordinator)
Natural Resource and Environment Programme
Thailand Development Research Institute Foundation
556 Ramkhamheang 39 Rd (Thepleela 1)
Wangthonglang
Bangkapee
Bangkok
Thailand
email: adis@leela1.tdri.ac.th

Zsuzsa Lehoczki (Eastern and Central Europe Region Coordinator)
Budapest University of Economic Sciences
Fovam Ter 8
1093 Budapest
Hungary
email: eapp@mail.matav.hu

Ronaldo Seroa da Motta
Instituto de Pesquisa Economica Aplicada
Av. Presidente Antonio Carlos 51
13e andar
CEP 20020–010
Rio de Janeiro, RJ
Brazil
Caixa Postal 2672
email: seroa@ipea.gov.br

Part I

Africa Case Studies

1

THE BOTSWANA TRIBAL GRAZING LAND POLICY: A PROPERTY RIGHTS STUDY

N H Fidzani[1]

BACKGROUND

Unlike in developed countries, where livestock stocking rates are determined by price changes, in Africa other factors, including the open access nature of rangelands, are more important. Indeed, the 'tragedy of the commons' perspective has been used to explain overstocking and the resultant range degradation and has influenced rangeland policies throughout Africa.

In environmental economics, the costs of natural resource use are broken down into direct costs, external costs and user costs of extraction. In this case, direct costs are those incurred by the herder when increasing his herd. External costs are those incurred by other herders as a consequence of one herder's act of increasing his herd. User costs constitute the opportunity cost to future users of resources arising from the current actions of an individual. The tragedy of the commons seeks to explain overstocking as a function of the divergence between the private and social costs of holding cattle. It argues that on deciding whether or not to increase his herd, the herder looks at the direct costs of doing so in relation to the benefit arising from such an action. His calculation does not include any external costs or user costs. Since he is going to pocket all the proceeds from the sale of an additionally-accumulated animal but gets to share its costs with society, the net benefit from such an expansion is always positive. Acting rationally, the herder would expand his herd up to a point where his marginal benefits from an extra head equals its marginal cost. The repercussion of this is overstocking and overgrazing. 'Tragedy of the commons' proponents further argue that the nature of property rights is such that the free-rider problem discourages any preservation and improvement of efficient resource use, culminating in range degradation.

Inspired by the need to eliminate the discrepancy between private benefits and social costs, much of the literature on the tragedy of the commons envisions the salvation of the commons to lie with their privatization. This, it is argued, will force individual herders to internalize the user costs and external costs, thereby compelling them to behave in a socially optimal manner (Cheung, 1970; Demsets, 1967; Posner, 1977).

The World Bank and bilateral agencies such as USAID and SIDA have fallen for this prescription and have sought to apply it to Eastern and Southern Africa. Botswana is one of the countries to which this prescription has been applied, through what has come to be known as the Tribal Grazing Land Policy (TGLP). This paper is an attempt to review the extent to which the TGLP, as an economic instrument, has succeeded in internalizing externalities through the privatization of the commons.

ENVIRONMENTAL ISSUES

The problem that was hoped to be solved by the TGLP is succinctly summarized by Botswana Government Paper No.2 of 1975:

> *Increased herds, under the system of uncontrolled grazing, have led to serious overgrazing around villages, surface water sources and boreholes. Overgrazing around villages has led to sheet erosion and bush encroachment which reduces the amount of grazing. This is worst for small cattle owners most of whose herds graze in the village.*

The paper continues to state that:

> *Under the present system, the wealthier cattle owners secure virtually exclusive rights to land around their boreholes. More and more grazing land gets taken up by few large cattle owners. Meanwhile those who own own a few livestock stay where they are in the villages with little hope of improvement.*

The above problem arose due to the rapid growth in the national herd as a result of increased use of veterinary medicine and a break-through in water technology which made available previously inaccessible rangelands, through borehole drilling. Consequently, the national cattle herd more than doubled in less than ten years from 1.2 million in 1966 to 2.3 million in 1974. It is important to note that most of this expansion took place in the ecologically delicate sand-veld ecosystem of the Kalahari Desert. A similar expansion also occurred in the hard-veld region where surface water was easily available – in this case the increase was fuelled mainly by government policies that made it extremely attractive to hold cattle.

Over-expansion, sheet erosion and bush encroachment were clearly threatening the livestock sector's long-term sustainability and economic viability. As ecological damage occurs, the ability of the rangelands to recover is reduced, and with it the land's long-term carrying capacity. It was also becoming clear that the growth of the sector was creating problems of inequity. To the extent that only those with the means and funds to drill boreholes were the ones whose herds were increasing, their expansion crowded out the small herders. This had a negative bearing on the social justice in the country.

THE TRIBAL GRAZING LAND POLICY AS AN ECONOMIC INSTRUMENT

As already mentioned, the TGLP was meant to address the common property problem arising from the communal use of rangelands: the benefits of overstocking accrue only to the farmer whose herd is increasing, while the resultant costs are shared by the whole society. These costs therefore constitute an externality to the society. To the extent that the benefits out-weigh the costs, the herder will always have an incentive to expand his herd, and the result will be overstocking. Another important aspect of a common property problem is the free-rider problem. Since a herder who invests in forage conservation or other rangeland improvement measures will not be able to retain all the benefits of such activities for himself, the free-rider problem arises as no herder will have an incentive to undertake such activities. This further complicates the management of a common property resource. The solution then lies in devising a property right arrangement that internalizes this externality.

The fact that the TGLP was inspired by the Tragedy of the Commons Argument is epitomized by the late President Sir Seretse Khama's TGLP inception speech:

> *Under our communal grazing system it is in no one individual's interest to limit the number of his animals. If one man takes his cattle off, someone else moves in. Unless livestock numbers are somehow tied to specific grazing areas, no one has an incentive to control grazing (Khama, 1975).*

In an effort to internalize this externality the TGLP then sought to privatize part of the rangelands. The stated objectives of the policy were to: (1) facilitate the introduction of improved management practices and range conservation; (2) increase productivity through the use of high off-take rates; and (3) to upgrade rural living standards by reducing the income gap between the rich and the poor. The reasoning behind the first objective was that exclusive rights over land were going to give farmers the incentive to control livestock numbers and improve management and range conservation. It was estimated that this was going to double the productivity of the rangelands through beef production. The narrowing of the gap between the rich and the poor was going to be achieved by moving the big herders from the concentrated area, thereby creating space for the poor to expand their herds.

Prior to the implementation of this policy, rangeland management in Botswana was the responsibility of traditional leaders such as chiefs. They were responsible for sanctioning which grazing lands could be used during specific seasons and could allocate specific areas to specific groups. However with increased modernization, these traditional institutions became weak. At independence in 1966 land control was handed over to the Land Board of the Local Authorities Department. These local Land Boards were charged not only with the responsibility of allocating grazing rights and water drilling rights but also allocating land for residential and arable purposes. As time went on and the national herd grew, there was pressure on these boards to allocate more water drilling rights. This was the beginning of the scramble for land in Botswana.

In seeking to address these problems, the TGLP subdivided the existing rangelands into three zones: Commercial, Communal and Reserved Zones, each of which was to meet the needs of particular groups of livestock owners.

Commercial Zones

These were the areas where private farms would be given exclusive property rights. Fifty-year renewable and inheritable leases were to be granted free to individuals or groups. It was hoped that these farms would be commercially managed using modern techniques such as water reticulation, fencing and high off-take rates. This group was also supposed to pay rent on these leases at a rate of P0.04 per hectare, or P250.00 (US$80.00) for an average-sized TGLP farm of 64 km^2.[2] The leaseholders were granted a grace period of five years during which no rent was to be paid. Revenue collected from these rents was supposed to be used to develop the communal zones.

The Communal Zones

These areas were to remain under communal use and this was where the majority of the small farmers would keep their cattle. Paragraph 40 of the 1975 White Paper, however, stipulates that each Land Board must determine and enforce a limit on the maximum number of animals which any individual or family could keep on a specific piece of communal land. It further stipulates that the construction and use of new and existing private water supplies should be restricted and new privately-owned boreholes would only be permitted if they were used for watering a few livestock and for arable purposes.

The Reserved Zones

These areas were to be reserved for future use by those with no cattle at the time and by future generations. These zones were also to be used for wildlife, mining and arable agriculture.

After the introduction of this policy the land in Botswana was allocated as follows (from Hubbard, 1986):

6%   Freehold farms
10% TGLP Commercial zone (700–900 ranches)
32% TGLP Communal zone
15% State Land (including forests)
11% TGLP Reserved Zone (wildlife management areas and areas for future livestock use)
26% land still to be allocated

It is clear from these data that according to this policy, about 42 per cent of the land was to remain in private hands (ie freehold, TGLP commercial farms and the area that was still to be allocated in 1986). Thirty-two per cent was to remain under communal use and the remainder was State land.

The change in land tenure as an economic instrument

This change in land tenure can be regarded as an economic instrument, since it internalizes the externalities that arise from the common property nature of rangelands. The land tenure change will therefore make it impossible for the herder to pass both his external and user costs on to society and he will have to face these costs alone. The second positive aspect of this change in tenure is that the herder will now have the incentive not only to conserve the rangelands, but also to adopt techniques to improve the land's productivity. The fact that he now has exclusive land rights means that he can plant grasses and fodder trees, knowing that no other herder will come to share them with him, thus eliminating the free rider problem.

Other economic instruments used with the policy

One of the requirements of the TGLP was that those who were allocated land were to pay a yearly rent for the use of these farms. This rental payment qualifies as a financial economic instrument, which is intended to complement the property right instrument. If applied properly, this rent should reflect the opportunity cost to the society of the lands that are now being exclusively occupied by the individual. This cost should include the benefits forgone by the whole society including those who have cattle and those who do not. The fact that there are some remote area dwellers who depend on this land for hunting and gathering means that their loss should also be reflected in this opportunity cost

The exclusive right holders were also supposed to receive interest-subsidised loans from the government through the National Development Bank, to be used for fencing and water reticulation. The easy access to these loans and the interest subsidy constitute additional economic instruments.

Another related economic instrument was the Drought Recovery Component of the Policy, which gave farmers access to a 60 per cent subsidy from the government for drilling and equipping boreholes. This was considered important because access to boreholes helps avoid the concentration of animals at single water points, which can lead to range degradation.

It is clear from the above discussion that a number of other economic instruments were used to complement the property right instrument, all of which were designed to encourage the proper utilization of range-lands.

The preparatory process
Once the government decided to go ahead with implementation of the policy, a major public education programme was launched, as consultation between the government and the people was considered fundamental to the successful implementation of the policy. The objectives of these consultations were to:

- provide information on the policy;
- stimulate public discussion;
- inform implementation boards, particularly Land Boards, how people felt the policy should be implemented; and
- inform people how they stood to benefit from the policy.

These consultations took the form of a nation-wide campaign by cabinet ministers addressing village-level public meetings; government officials leading seminars; and the formation of radio learning groups at village level.

Evaluation of the TGLP
The most comprehensive evaluation of this policy has been done by Carl Bro International (1982). As summarized in Table 1.1, this work reveals that the objective of doubling the productivity of rangelands has not been achieved. Subsequent work by other scholars has also confirmed this.

Table 1.1 *A Comparison of Productivity between TGLP Farms and Communal Areas.*

	Area 1		Area 2	
	Samane (Privatized)	Maikane (Communal)	Haina Veld (Privatized)	Tsau (Communal)
Total Output	8861	1234	9079	1795
Total Cost	3645	241	1951	249
Margins	5216	993	7128	1546
Margin/Head	27.4	26.4	17.3	18.1
Margin/Cow	44.8	57.4	42.5	43.5
Total Capital	42,246	4495	48,208	6966
Investment	12.4	22.1	14.7	22.2
Margin/P100	59	52	54	48
Calving Rate %	8.1	8.8	3.4	9.3
Death Rate%	9.3	6.2	10.3	12.5
Off Take Rate %	13.9	7.9	4.3	8.6

Source: Carl Bro International (1984)

The table reveals that:

- Within the same geographical area, output per head is approximately the same for the privatized ranches as for communal ranches. This, as argued by Carl Bro, indicates that there exists little difference in the performance level of the two systems.
- In terms of costs per head, Carl Bro found that these were higher for the privatized TGLP ranches than for the communal areas. This was mainly due to the fuel expenses, repairs and fencing maintenance costs incurred by ranches – which are not part of communal area expenses.
- Margins per head and per cow were found to be lower on the privatized ranches than in communal herds.
- The return on capital was found to be 22 per cent on communal herds and 13.5 per cent on ranches.
- Technical coefficients such as calving rates, mortality rates, off-take rates do not indicate any significant difference between the TGLP farms and communal areas.

These results cast some doubt on the validity of the policy's initial assumption that cattle production technology would enable many more cattle on the same amount of land on privatized rangeland. According to Hubbard (1986), this assumption was made on the basis of uncosted technical research results. In reality, the rate of return on the fenced farm investments emerged as unattractive. This concurs well with and explains Tsimako's (1991) finding that only 47.2 per cent of allocated ranches have been fenced, as can be seen from Table 1.2.

Table 1.2 *Progress with the Development of Allocated Ranches*

	Number of Ranches	Fence Completed	Fence Partially Completed	No Fence
Ngamiland	87	53	3	31
Central	73	21	7	41
Ngwaketse	52	36	7	9
Kweneng	37	20	0	17
Kgalagadi	31	4	2	25
Ghanzi	6	1	3	2
Total	286	135	22	125
%		47.2	7.7	43.7

Source: Tsimako (1991).

Several explanations have been given for this discouragingly poor performance of the TGLP farms. Empirical research has found that the granting of exclusive rights has not resulted in better range management on most ranches and off-take rates have remained low mainly because of the mismanagement of these farms. Cattle are often watered from single water points, and no water reticulation, fire breaks or paddocks have been created. Most ranches have not used disease or parasite control methods, maintained appropriate bull–cow ratios or practised forced weaning. Few farms have employed qualified ranch managers. All these shortcomings have resulted in overstocking, overgrazing, inefficient utilization of rangelands and poor performance.

Perimeter fencing, where it has been done, has been with a view to ensuring exclusive possession of the grazing land and preventing the cattle from straying off the land, as much as it has been done to improve the management of the rangeland. According to Behnke (1984) this should not be surprising since, historically, land privatization in Africa has always been for the purpose of controlling it rather than properly utilizing it. Behnke argues that in East Africa for example, participation in land privatization was motivated by the need to reduce the competition for arable purposes rather than increase its productivity. He further argues that it is for this reason that only those with strong political influence and economic power participated in these programmes – an aspect that can also be seen with the TGLP.

Because of the influential nature of the TGLP farm holders, they have been able to move their cattle temporarily off their ranches into the communal areas if their ranches become overgrazed or burnt by veld fires, or if their boreholes have broken down. This enables the ranchers to let their land recover. Other ranchers choose to leave some of their cattle in the communal areas as insurance against total loss, in case drought, fire, diseases or other disasters hit. In this way, TGLP farmers benefit disproportionately since, in addition to having exclusive property rights on large TGLP ranches, they also have access to communal grazing. By enjoying dual rights, ranchers avoid the consequences of overstocking and overgrazing their own ranches and this has acted as a disincentive to their adopting improved management practices. Dual rights and the ability of TGLP farmers to keep their cattle on the communal areas has led to an increased strain and overstocking on communal rangeland which has led to increased range degradation. Dual rights have also exacerbated the problem of inequity that the policy was meant to address. By holding some of their cattle on communal lands the farmers have been able to crowd out small farmers.

To the extent that the rich and politically influential are normally urban based, many TGLP farms have been held under absentee management. This has made it difficult for the ranchers to receive extension advice and to effectively oversee the necessary technical and management operations. The herders who manage these ranches lack the necessary training, guidance and skills required for effective ranch management (Tsimako, 1991). This has happened in spite of the clear stipulation by the lease that in cases where the owner cannot be fully based on his farm, a trained farm manager must be employed. Power and influence have been used to ignore this requirement.

Design and Implementation Problems

The failure of the TGLP can be largely attributed to the fact that false assumptions were made during its formulation. Hubbard (1986) has made an effort to analyse these assumptions and the first mistake he has identified is that it was initially thought that there was enough land to go around and so some land could be set aside for future generations as reserved areas. He points out that when the policy came to be implemented, no land was set aside as reserved areas, due to the fact that the amount of available land had been overestimated. According to him, much of the land was found to be unsuitable or already populated by hunters and gatherers. This has actually resulted in a legal battle as to whether these nomads have legal rights to the land in its totality or just for the purposes of hunting. The then Attorney General's view was that they do not have any land rights.

Hubbard summarizes another serious miscalculation that occurred during the policy's formulation:

Rather surprisingly, the confident prediction of the 1975 White Paper regarding raising of cattle productivity was made only on the basis of uncosted technical research results. When these technical coefficients relating to productivity differences between cattle posts and minimally fenced farm ranches were applied to budgets, the rate of return on the fenced farm investment emerged as unattractive even on the basis of the unrealistic assumption that ranch-type

*productivity levels are achieved immediately the investment is
made.*

Another false assumption was that it would be politically feasible to
apply compulsory stock regulations in both commercial and communal
areas. The White Paper failed to state how this was to be achieved. This
gave way to the interests of those with political power to practise dual
land rights.

The failure of these assumptions to hold, coupled with a lack of political will to implement some of the requirements, led to problems in the
implementation of the policy.

SUMMARY AND CONCLUSION

This case study has established that the privatization of rangelands
through the TGLP has not been very successful in bringing about
increased productivity. It has further established that the privatization
has brought about increased inequity by maintaining dual land rights for
a few influential people. The study has also found that even though there
has been some attempt to charge rent for the exclusive rights, the charges
levied were too low.

The problems of low productivity and increased inequality can be
explained in terms of the self-interest of the policy designers. Those who
took over private rangeland did not do so to improve livestock production but simply to have control over the land. Some see land as their
security for the future when they retire, some hold it for purely specula-
tive reasons. It is obvious that as both the human and livestock
population grow, there is going to be increased competition for land.
This means that land, particularly privatized land, will sell for a high price
in the future. This has no doubt led to the current misuse of land.

It is interesting to note that while this policy was supposed to
enhance the definition of property rights, it actually clouded it further
by allowing dual rights. It is these dual rights arrangements that have led
to increased inequality in livestock distribution. This has occurred
through the fact that those farmers with exclusive rights to specific
pieces of land have continued to have access to the commons. In a way,
this has exacerbated the rangeland degradation problem in that it has led
to a more inefficient use of land.

The policy also suffered from an improper valuation of the natural
resources. Clearly, the five-year grace period on rent payment, combined
with the subsequent rent payment of only about US$80 per annum for a
64 km^2 ranch is too generous and grossly underestimates the value of
these lands. While the carrying capacity of these ranches is at least 300
cattle and the average price of a fully-grown animal is about US$300, the
US$80 payment for the whole farm for a full year is a pittance. There is

no doubt that this has attracted the participation of people who are not even interested in rearing livestock. The major problem with this rental rate is its failure to take into account the welfare lost by those remote area dwellers (bushmen) and wildlife that have been displaced by the establishment of these ranches. A proper valuation taking this into account would have increased rents at least a hundred-fold.

The final question to address is whether with this experience of the TGLP, we can still claim that privatization is the answer to range degradation. The answer seems to be no. The tragedy of the commons literature has been criticized both on empirical and theoretical grounds. A tragedy of the commons perspective suggests that even when collective agreements are made *not* to expand herds, herders will always have the incentive to renege or defect, thereby producing an unstable equilibrium. This, it is argued, suggests the need to impose authority from outside the system to make the agreement work.

Perrings et al (1989) reject this argument on the grounds that it is ahistorical in that in-built institutions have always existed to either directly or indirectly handle the problem of overgrazing. They cite the Botswana case in which water syndicates have been found to have de facto control on rangelands without privatizing them.

Swallow et al (1984) have produced similar evidence from Lesotho that indicates that traditional appointees are assigned the responsibility of controlling and regulating grazing. Similar institutional arrangements are said to exist in East Africa. It has also been argued that privatization of rangelands actually overlooks the unpredictability of the spatial distribution of rainfall in this region. The only security against this unpredictability is the ability to move herds from one place to another – something that can not be done under the enclosure system.

NOTES

1   University of Botswana.
2   P = Pula; approximately P3.85 = US$1.

REFERENCES

Behnke, R H (1984) Fenced and Opening – Ranching; The Commercialization of Pastoral Land and Livestock in Africa, in Simpson J R and Evanglon P, *Livestock Development In Africa: Constraints, Prospects and Policy*. West View Press, Boulder.

Cliffe, L, and Moorson, R (1979) Rural Class Formation and Ecological Collapse in Botswana. *Review of Africa Political Economy* vol 15/16.

Carl Bro International (1984) An Evaluation of Livestock Management and Production with Special References to Communal Areas. Carl Bro International, Denmark.

Cheung, S N S (1970) The Structure of a Contract and the Theory of a Non-exclusive Resource. *Journal of Law And Economics* vol 13: 49–70.

Demsets, H (1967) Towards a Theory of Property Rights. *American Economic Review* vol 57: 347–359.

Gordon, H (1954) Theory of Common Property Resource. *Journal of Political Economy* vol 62; 124–42.

Hardin, G J The Tragedy of the Commons. *Science* vol 62: 1243–1248.

Hubbard, M (1986) *Agricultural Exports and Economic Growth: A Study of the Botswana Beef Industry.* KPI Limited, London.

Khama Seretse (1975) National Policy on Tribal Grazing Land. Government White Paper no 2, Botswana Government Printer, Gaborone.

Larson, B A and Bromley, D W (1990) Property Externality, and Resource Degradation: Locating the Tragedy. *Journal of Development Economics* vol 33 (2): 235–261.

Parson, J (1979) The Political Economy of Botswana: A Case Study of Politics and Social Change in Post-Colonial Societies. D Phil Thesis, University of Sussex.

Perrings, C, Opschoor, J B, Arntzen, J W, Gilbert, A, and Pearce, D W (1988) Economics of Sustainable Development: Botswana, A Case Study. Technical Report prepared for National Conservation Strategy, Gaborone.

Picard, L (1980) Bureaucrats, Cattle and Public Policy: Land Tenure Changes in Botswana. *Comparative Political Studies* vol 13 (3): 313–356.

Posner, R A (1980) A Theory of Primitive Society, with Special Reference to Law. *Journal of Law and Economics* vol 23: 1–53.

Runge, C F (1981) Common Property Externality; Isolation, Assurance, and Resource Depletion in Traditional Context. *American Journal of Agricultural Economics* vol 63: 596–606.

Swallow, B, Mokitini, B, and Brokken, R (1984) Cattle Marketing in Lesotho. Research Report no 13, Institute of Southern African Studies, National University of Lesotho.

Tsimako, B (1991) The Tribal Grazing Land Policy (TGLP): Performance to Date. Botswana Ministry of Agriculture, Gaborone.

2

THE MANAGEMENT OF FOREST
RESERVES IN BOTSWANA

N H Fidzani[1]

BACKGROUND

Forests in Africa, as on all other continents, contribute to food security by providing hunting and gathering opportunities, and provide energy from fuelwood, as well as providing environmental protection through watershed stabilization, soil protection and carbon sequestration. Despite these important roles played by forests, the manner in which forests have been exploited in the Sub-Saharan region is not encouraging. While the extent of deforestation in this region as a whole is estimated at 1 per cent, below the world average, some parts of the region have alarming deforestation rates. It is estimated, for example, that 75 per cent of Western African forests have been lost (Sharma et al, 1994).

Deforestation in Africa is a consequence of a number of factors. The most quoted cause is high human population growth, which creates increased pressure to open up more arable land for farming, at the expense of forests. Another important factor is the need to acquire foreign exchange through international trade. Sub-Saharan African countries need to earn foreign currency in order to be able to sponsor development projects and the need to earn this capital has, in some cases, been given higher priority than has the conservation of forests and other natural resources.

Deforestation can also be considered a consequence of the institutional problems associated with the common property nature of forests. This has made it difficult for markets to operate efficiently in this sector and has led to the undervaluation and underpricing of forest products. However, there has been a growing awareness among most African governments that sustainable economic growth can only be achieved through the proper use and conservation of their natural resources. This growing awareness has been accompanied by several efforts to use market forces and internalize any externalities to try and achieve optimal resource use in this sector.

This chapter analyses how forests have been managed in Botswana, to demonstrate how economic instruments have been used in this sector. The chapter is based on a case study undertaken by the Norwegian Forestry Society (NFS), on behalf of the Ministry of Agriculture, on the forests in the Chobe District of Botswana, as well a separate study by Kgathi on the use of firewood in Botswana.

LEGAL CONTEXT

The Botswana Forest Act of 1968, which was amended in 1976, confers powers to the President of Botswana to establish forest reserves on state land. As pointed out by Kgathi (1992), the act also permits local authorities to request the minister concerned to establish forest reserves under their remit. Once these areas have been declared forest reserves, the cutting and burning of trees within their boundaries is prohibited without prior permission. Furthermore, the act grants local authorities powers to introduce by-laws for charging and collecting fees and royalties. The act also gives the minister the power to declare certain trees protected, in any part of the country.

Another legal instrument for controlling the use of forests is the Agricultural Resources Act of 1974 whose objective was to conserve agricultural plants, animals and vegetation. An Agricultural Resource Board with ten different committees was formed to issue conservation and animal stock control orders, develop regulations on conservation and advise local Land Boards on land use. It is this last aspect that is relevant to forest management.

The third legal instrument which has had a bearing on forest utilization is the Town and Regional Planning Act of 1977, which gives the minister the power to establish any area in Botswana as a planning area. According to Kgathi (1992), restrictions can be placed on the cutting of trees in planning areas, and so this act, like the other legal instruments, has influenced the utilization of the country's forests.

Botswana has therefore combined command-and-control policies and economic instruments to regulate and influence the management of its forests. For example, the provision by the Forest Act that the minister can declare certain trees protected is a command policy. However, the Forest Act also provides for the use of economic instruments such as fees and royalties. The following is an analysis of the extent to which these economic instruments have been successful in conserving the Chobe Forests which were declared forest reserves by the President of Botswana under the purview of the Forest Act.

THE MANAGEMENT OF THE CHOBE FORESTS

Forests in the Chobe district of Botswana are dominated by teak or *Baikia plurijuga* (local name *mukusi*, also known in Botswana as teak), *Pterocarpus angolensis* (or *mukwa*) and *Burkea africana*. *Mukwa* is the most economically important species, although the forests are actually dominated by *mukusi*. This part of Botswana is inhabited by elephants, which cause considerable damage to the forests, both directly by pushing down trees and indirectly, by damaging trees and thereby making them less tolerant of fire. *Mukwa* has suffered most from the elephants (an estimated 18 per cent of all *mukwa* trees had been pulled down by elephants at the time of the study) and this species also seems to be affected by a disease that is spreading in the district. The domination of the forest by the fire sensitive *mukusi* and the existence of large populations of elephants have made fire the most serious threat to these forests. It has been estimated that 55 per cent of the first year-old shoots of all species are killed by fire. The combined effects of fire and elephant damage have resulted in a low number of *mukwa* stems of less than 35cm diameter. This seriously threatens the long-term sustainability of these forests as there will be less young trees to replace the old dying stock. A conservation-oriented approach to cutting these forests would extract the damaged trees and leave the healthy ones. The following analysis attempts to demonstrate how the use of the economic instruments has not only failed to achieve this objective, but has actually exacerbated the problem.

IMPLEMENTATION ISSUES

Royalties and Stumpage Fees Levels

The NFS report states that:

> In 1990, the government earned a total of P131,046 in royalties from logging operations in the Chobe District.[2] Income from logging is low because stumpage fees bear little relationship to the true economic or market value of the timber resource. These royalty rates are much lower than in neighbouring countries, where the system of setting fees is influenced by market forces.

The NFS financial analysis reveals that the Botswana government lost potential revenues of about P2.46 million for 1990, as a consequence of the generous stipulation that stumpage fee payment should be made only on the parts of the harvested trees greater than 35cm diameter. Logging companies therefore pay the same flat fee for very large trees as for trees of only 35cm diameter. In 1990 the total harvest volume of *mukwa* registered was 8065m³, yet royalty was paid on only 3302.6m³. Another 568.7m³ was taken free as it fell below the minimum diameter and an estimated 4194m³ was left to waste on the forest floor. Similarly, for teak a total of 5159m³ was harvested of which 2173.7m³ attracted royalty, 302.7m³ was taken free and 2682m³ was estimated to be left to waste.

A number of inefficiencies can be identified from the above situation. First, as clearly stated by NFS:

> *The logging concessionaires are transferring economic rent from the district to themselves because the price paid for timber resources is far below its true value. Low stumpage fees have resulted in wasteful utilization of scarce timber resources. The concessionaires receive an indirect subsidy from the government because the costs of managing the resource and monitoring the logging operations are not covered by stumpage fees.*

These low stumpage fees have also been responsible for an increased fire hazard in the forest (from the harvested trees left on the forest floor) and exploitative, rather than conservative, harvesting methods (ie creaming off the biggest healthiest trees instead of salvaging the damaged ones).

Fire Hazards

As the NFS stated, fire damage has increased significantly in the logging areas as significant amounts of harvesting residues are left near living trees and these residues then ignite in subsequent fire seasons. In this regard, low royalty charges have exacerbated the fire problem that is faced by the Chobe forests.

Harvest By Creaming-Off Instead Of Salvaging

As already mentioned, the damage from fire and elephants would seem to suggest that harvesting these forests should focus on salvaging the damaged trees rather than felling the healthy ones. The low and non-differential stumpage fees have instead encouraged the creaming-off of healthy trees, leaving behind damaged ones. This shows that even though economic instruments have been used to try and influence forest harvesting, they have not been used in a proper and effective manner. As will be shown later, encouraging salvaging could have been achieved by differential stumpage fees, whereby high fees are charged for fresh trees and low fees for damaged ones.

NFS has estimated that damaged stock alone constituted up to ten years of harvest at current offtake rates. Another inefficiency loss emanating from this stumpage fee system has been heavy damage to trees from logging and extraction. To illustrate the extent and seriousness of this inefficiency, the NFS study quotes a mill recovery rate of 18 per cent, which by their judgement is very low.[3]

Inefficient Use of Teak

An additional inefficiency in the harvesting of trees from the Chobe Forests relates to teak. The local market for this species is limited and its main export market is South Africa, whose growing sophistication demands high quality hard wood. This has made Botswana compete for this market with the USA, South East Asia, and currency-strapped African exporting countries such as Mozambique, Zaire and Zambia. The prices fetched by this species have been said to be very low. The Botswana teak harvest is turned into three main products: South African mine sleepers (60 per cent), wide boards (20 per cent), and parquet floor batons (20 per cent). The study by NFS has found that the processing of this product is very inefficient, with a mill recovery rate of 37 per cent. When the losses in the field are added to the losses in milling, the recovery rate falls to 18 per cent. NFS stated that:

> Milling costs are high and output is low due to the combined effects
> of inefficient and outdated equipment, and low productivity.
> Inadequate log haulage, transport, and poor management are other
> factors. Milling costs average P691 per m³ (excluding felling,
> hauling, transport and royalty) whilst the average price for the tree
> products is P608 per m³.

This suggests that the processing of teak is unprofitable. This finding prompted NFS to conclude that if a high value market for this species cannot be found, it would be prudent to leave the trees standing until such markets are developed. They further recommended that the private sector be encouraged to develop such markets, by setting the royalty at a level which would make it unprofitable to cut and turn teak timber into lower-priced South African mine sleepers. They also recommended imposing an export tax on logs.

To sum up, low royalties have enabled timber harvesting to occur even when market prices are depressed and high inefficiencies in processing prevail. This is a clear case of wastage in the utilization of forest products, caused by undervaluing and underpricing these forest products.

To a certain extent, the non-sustainable exploitation of forests can also be explained by the short-term nature of the concessions issued. The NFS report states that the main company exploiting these forests' resources, PG, has taken a short-term investment strategy. They state that:

*PG's objectives were to maximize extraction, to ensure quick
recovery of investments and make a reasonable profit. By the end of
1991, PG had achieved its objectives in all areas. The company
appears to have taken a short-term view because of uncertainties
over the concession beyond February 1993.*

This confirms that undefined and uncertain property rights can lead to
the overexploitation of natural resources.

BOTSWANA'S CURRENT STRATEGY TO ADDRESS FOREST POLICY FAILURES

In response to the above problems, the Botswana Government accepted
NFS's recommendation that substantially higher royalty fees be introduced to eliminate the hidden subsidy from the government to logging
companies. These are shown in Table 2.1. These high fees are expected to
discourage companies from logging teak and to encourage them to search
for a high-priced market for this species. The stumpage fees structure has
also been revised to include management fees that are levied to cover the
cost of monitoring these companies by the government. Silvicultural fees
have also been introduced to establish a fund for fire protection and the
construction of fire breaks. To encourage the cleaning up of tree branches
after logging, heavy penalties have been imposed on those who do not
comply. Furthermore, to encourage salvage harvesting and discourage
creaming off, a differentiated pricing system has been introduced, charging a higher royalty for healthy trees and a lower one for damaged stock.
To discourage wasteful harvesting, the royalty structure was changed to
cover the whole tree (ie the stem and branches) instead of just the stem as
was the case before. This has been done with a view of encouraging
logging firms to clean up after their logging, to avoid fire hazards.
    The change in royalty structure has been reinforced by some insti-
tutional changes. It has been decided that instead of having many
small logging companies that are too difficult for the Forestry
Department to supervise, only a few large companies are to be issued
concessions. Furthermore the Forestry Department itself will need
strengthening before all these new policies can be implemented. The
government has therefore frozen all logging activities from 1993 to
date, pending the outcome of a forestation policy, which is currently
being developed. It is expected that this policy will be finalized before
the end of 2000.

CONCLUSION

This case study has shown that whilst an attempt has been made to use economic instruments such as stumpage fees to control the exploitation of forest resources, this attempt has not been very successful due to a number of inefficiencies. The problem seems to lie with the policy makers' failure to set these instruments at levels commensurate with their opportunity cost. Botswana could have improved the efficacy of these instruments by using appropriate economic techniques to properly value the forest resources. There is therefore a need to apply these valuation methods before economic instruments are set.

Table 2.1 *The Old And New Royalty Structures*

	Old structure	New structure
1 Management Fees per m^3	None	P5.25
2 Silvicultural Fees per m^3	None	P3.00
3 Undamaged Mukwa per m^3	P39.68	P210.00
4 Damaged Mukwa per m^3	P39.68	P157.50
5 Undamaged Mukusi per m^3	P19.84	P157.50
6 Damaged Mukusi per m^3	P19.84	P105.00

Source: Ministry of Agriculture (interviews)

NOTES

1   University of Botswana.
2   P = Pula; approximately P3.85 = US$1.
3   The mill recovery rate refers to the percentage of the log that is transformed into the end product. A mill recovery rate of 18 per cent means that 18 per cent of the log that enters the machine ends up being usable. 82 per cent is thrown away as waste.

REFERENCES

Erkkila, A and Siislonem, H (1993) *Forestry in Namibia 1950–1990*. University of Joensuii, Finland.

Kgathi, D L (1992) Household Responses to Fuelwood Scarcity in South-Eastern Botswana: Implications for Energy Policy. PhD thesis submitted to the School of Development Studies, University of East Anglia.

Krames, R A N, Sharma, H, and Munasinghe, M (1995) Valuing Tropical Forests: Methodology and Case Study of Madagascar. World Bank Environment Paper no 13, The World Bank, Washington, DC.

Harrilton, A C (1984) *Deforestation in Uganda*. Oxford University Press, Nairobi.

Norwegian Forestry Society (1993) Chobe Forests Inventory and Management Plan. A Technical Report prepared for the Ministry of Agriculture, Division of Crop Production and Forestry, Government of Botswana.

Sharma, N P, Rietbergen, S, Heimo, C R and Patel, J (1994) A Strategy for the Forest Sector in Sub-Saharan Africa. A World Bank Technical Paper no 251, The World Bank, Washington DC.

3

PROMOTING A SUSTAINABLE WATER SUPPLY IN BOTSWANA

J W Arntzen[1]

BACKGROUND

Water scarcity is expected to become a major issue in the next century on a global and regional level as well as on a national level for Botswana (see for example Agenda 21 (United Nations 1992) and Brown et al, 1996). Without an expansion of Botswana's water supply systems, water shortages are expected by the late 1990s or early next century. The geographical divide between areas of increasing water demand and areas offering additional water supply opportunities will necessitate the construction of a north-south water conduit system and the implementation of demand control strategies. And, as poverty is still widespread in Botswana, water supply systems need to be based on a balance of efficiency, sustainability and social justice. Demand predictions for the country are given in Table 3.1.

Table 3.1 *Expected Changes in Water Demands in Botswana (1990–2020, million m³)*

Category	Water Demand 1990	Water Demand 2000	Water Demand 2020
Settlements	33.8	68.9	167.8
Mining and Energy	22.9	33.6	58.9
Livestock	35.3	44.8	44.1
Irrigation/Forestry	18.9	30.2	46.9
Wildlife	6.0	6.0	6.0
Total demand	116.9	183.5	323.6

Source: SMEC, WLPU, SGAB, 1991.

The water supply problem may be compounded by the effects of global warming, which would increase evapotranspiration and reduce the lifetime of the (mostly shallow) dams. Most countries in this region have experienced rapidly rising water supply costs, and have had to turn to expensive water transport systems (including Botswana, Namibia, and Zimbabwe). South Africa has even accessed water from neighbouring countries (including via the Lesotho Highland Water Scheme). Moreover, pressure is mounting on shared water resources such as the Zambezi, the Limpopo and the Okavango systems.

Growing water consumption levels are the result of several factors:

• rapid population growth (3.5 per cent per annum between 1981–1991);
• increasing income levels leading to higher per capita water consumption (per capita income level was US$ 2 790 in 1992);
• rapid economic growth; and
• uneven population distribution: around 80 per cent of the population live in eastern Botswana, and half the population now live within a 100km radius of the capital, Gaborone (in south-eastern Botswana).

Water supply systems are increasingly stretched because of:

• low recharge of aquifers (an average rate of 3 mm per annum);
• erratic rainfall patterns (increasing from 250 to 600 mm from the southwest to the north);
• few shallow dam sites; and
• high evapotranspiration (2000 mm per annum).

Various institutions are responsible for the country's water supply: the Department of Water Affairs (DWA) and District Councils (in large villages), Water Utilities Corporation (WUC, in urban areas) and individual users (including mines and borehole owners). The country has a dual supply system. On the one hand, DWA, district councils and WUC supply urban areas, including most of the industries and large villages, at a monthly supply charge according to the amount of water consumed. On the other hand, consumers in remote areas, livestock owners and some of the mines have to secure their own water supply. These users need to cover all the costs themselves, though they are not necessarily subject to a charge related to the consumption level.

WATER USER CHARGE

The principal economic instrument used in Botswana to manage water use is a differential user charge. A secondary instrument takes the form of a subsidy to make potable water affordable to the entire population. The equation for water charges through the Water Utilities Corporation is:

production costs + transport costs – subsidies

Because of water scarcity, the production costs and transport costs are increasing and at the same time the government intends to reduce water subsidies. Therefore, real water prices are expected to increase substantially and prices will better reflect the user-pays principle (excluding external costs and forgone benefits). Current differences in charges are based on two factors:

1   *location:* water charges for users near large supply sources are lower (due to lower transport costs); rural water users receive more subsidy (and will continue to do so), and the overall charges in rural areas tend to be lower than in urban areas. Charges are highest in Gaborone (see Table 3.2).
2   *level of consumption:* no difference is made between domestic and commercial use but, for social reasons, the charge levied on the first 10 m^3 of water consumed is relatively low, and higher consumption is 'stepwise punished' through increased unit charges (see Table 3.2).

The long run marginal supply costs (LRMC) are estimated at US\$ 1.94/m^3 in Gaborone and US\$0.62/m^3 in Francistown (at 1990 prices). Again, these costs exclude environmental externalities and forgone benefits of future users.

The expectation is that water charges will reduce the per capita water consumption rates, and therefore alleviate water scarcity. Realizing that subsidies have the opposite impact, it is now policy to reduce subsidies and to target the remaining subsidies to those most in need, for reasons of social justice.

Table 3.2 *Water Charges in Gaborone and Large Rural Villages;*
(US\$/month)

Water consumption band	Gaborone 1993	Gaborone 1996	Large villages 1993
first 5 m^3	35	39	18
next 5 m^3	35	39	37
next 5 m^3	102	114	37
next 5 m^3	131	146	37
next 5 m^3	131	146	73
next 5 m^3	179	201	73
next 5 m^3	179	201	73
over 40 m^3	179	201	163[a]

Note: a only applies to two semi-urban villages.
Sources: Arntzen, 1995, p.339; 1996/ 97 Water Utilities Corporation

Water charges were in fact considered as soon as water scarcity became a national issue back in the 1980s. During the 1980s drought charges were increased and the National Water Master Plan, prepared in the late 1980s, provided the required background for the implementation of higher water charges. During the 1980s drought, voluntary water restrictions proved highly successful in reducing short run water consumption. This was achieved through awareness campaigns using radio and newspaper advertisements.

A number of regulatory instruments control water consumption, directly or indirectly. Granting of water use licences has been used to control the location of industries. The regulation that boreholes be at least 8 km apart restricts the number of boreholes sunk (although there is no restriction on the water consumption per borehole).

Water charges for all urban users and most industrial users are set by the Water Utilities Corporation, and require government approval. In large settlements, district councils and the Department of Water Affairs jointly determine the charges. The same institutions are responsible for collecting the revenues. Water charges are annually reviewed.

DESIGN AND IMPLEMENTATION ISSUES

No distinction is made between productive and consumptive use. Therefore although the rates are incremental, there is no added incentive for companies to reduce water consumption, because they cannot escape the highest tariff band.

In low-income areas, standpipes are the main source of water. In villages, standpipe water consumption is free; in urban areas a flat monthly service rate is charged for basic government services, including water supply. Consequently, there are no economic incentives to curb water consumption.

The instrument only covers 'market consumption' in urban areas and rural settlements. The livestock sector, the single largest water consumer, is not affected by the charges.

Botswana's water charges are considerably higher than charges levied in neighbouring countries. This has caused frequent complaints from the private sector, and is sometimes cited as a disincentive for companies considering investing in the country. Because of soaring water supply costs, neighbouring countries will have no choice but to increase water charges, and therefore the price difference is likely to be reduced in the future.

Some peri-urban areas close to Gaborone receive water from urban supplies at high costs, but sell it at lower rural charges (benefiting from the government subsidies). This appears unjustified where the water is being used for lawns, swimming pools and other forms of luxury consumption. Another problem observed is the relatively expensive system for collecting the charge revenue in rural areas.

IMPACTS

Water charges have probably contributed to the relatively low per capita water consumption, but their impact is probably more limited than was originally envisioned. This is due in part to the continued subsidization of water supply, which in effect lowers the charges. It is well known from the literature that charges are more effective when they are high (Opschoor and Vos, 1989). Another indication of the limited impact can be found in the fact that despite the lower water charges in the north, few companies have relocated there. The low consumption figure is also determined by other, possibly more important, factors such as:

- income inequality and the prevalence of low incomes; for example, daily water consumption in urban low income areas ranges from 28 to 65 litres per person compared with 346 to 382 litres in urban high income areas. In rural areas, the average per capita consumption in houses with running water is ten times the per capita consumption in households that depend on village standpipes.
- economic structure: the absence of irrigation and the lack of a large manufacturing base has kept commercial water consumption relatively low.
- inadequate access to water in remote areas (whilst a diminishing problem) has also suppressed water consumption.

High water charges have had two positive resource use impacts. Firstly, high water charges in conjunction with the government's unwillingness to subsidize irrigation have kept the irrigation sector small. This is in clear contrast with neighbouring South Africa and Zimbabwe, which both face the difficult task of achieving a more efficient water allocation system, including charging for irrigation without immediately killing the sector. Both countries are now revising their irrigation strategies and are likely to bring them more into line with Botswana. Secondly, high water charges provide incentives for water-harvesting and water-saving practices and encourage reductions in the substantial losses in existing reticulation systems (where water loss can be as high as 20 per cent of the water volume in the system). There has been no assessment of the extent to which such water-economizing measures have been done in practice.

Higher charges have undoubtedly contributed towards a greater degree of cost recovery within the water sector, but subsidies still remain high. Increased cost recovery will release government finances for other development sectors. Because of the minimal charges for low consumption levels, the policy has been successful in promoting social justice.

CONCLUSIONS

Given the continued subsidies, water charges are not yet efficient. However, their efficiency is increasing with the reduction and targeting of subsidies. There are several possibilities for further increasing the efficiency, including merging the charge bands for urban and rural areas. Those rural people who can afford private water connections should be exposed to the same charge bands as urban-based users. Moreover, subsidies for water-saving devices and technologies could be considered in cases where there are some net private costs but net social benefits. Finally, incentives should be given to promote productive water use. For example, productive water consumption should be charged at lower rates than luxury domestic consumption such as watering lawns.

Water supply seems to be fair from an equity perspective. It is important that the reduction in subsidies should not affect the lowest income groups. Water consumers already perceive water charges as very high and higher charges would no doubt receive a hostile response from users. The administrative costs of the water charge are relatively low in areas with many private connections, though in rural areas these costs, including penalty procedures and reconnections, must be substantial. The water charges, and high utility costs in general, are said to be scaring off foreign investors, although there is reason to be sceptical about this argument. For most companies, water costs are only a minor consideration in their decision on where to locate their business. Market factors, including the availability and costs of labour, seem to be much more important. This has been demonstrated by the failure of the relatively low water charges levied in the north to attract more companies to relocate there.

In order to ensure sustainability of the nation's water resources, a comprehensive policy should be prepared covering *all* uses and users. At present, the non-market water sector is not covered, and the livestock sector, for instance, has no incentive to use water efficiently. It is therefore necessary that all institutions responsible for supplying water coordinate their policies and actions.

A combination of instruments, legislative, economic and consultative ones, appear to produce the best results. For example, awareness campaigns and voluntary restrictions proved effective during the 1980s drought. Such a combination of instruments could also be used to safeguard the water quality. In the absence of water pollution control policy, licences have proved effective in preventing companies from undertaking polluting activities in catchment areas.

NOTE

1   University of Botswana

REFERENCES

Arntzen, J W (1995) Economic Instruments for Sustainable Resource Management: the Case of Botswana's Water Resources. *Ambio* vol 24 (6): 335–342.

Ayub, M A and Kuffner, U (1994) Water Management in the Maghreb Region. *Finance and Development* June 1994: 28–29, The World Bank, Washington, DC.

Brown, L R et al (1996) *State of the World 1996.* Norton/Worldwatch Institute, New York.

Carruthers, I and Stoner, R (1981) Economic Aspects and Policy Issues in Groundwater Development. World Bank Staff Working Paper no 496, The World Bank, Washington DC.

Dixon, J, Talbot, L, and le Moigne, G (1989) Dams and the Environment. World Bank Technical Paper no 110, The World Bank Washington DC.

Hassan, R, Breen, C and Mirrilees, R (1996) Management of Water Resources and Emerging Water Policy Challenges in South Africa. Paper presented at the Workshop on Economics, Policy and Natural Resource Management, CSIR, South Africa.

Khupe, B B (1994) Integrated Water Resource Management in Botswana. In: Gieske, A and Gould, J (eds) *Integrated Water Resource Management: Workshop Proceedings.*

Lundqvist, J and Jonch-Clausen, T (eds) (1994) Putting Dublin/Agenda 21 into Practice: Lessons and New Approaches in Water and Land Management. Special Session at VIIIth IWRA WorlDCongress, Cairo.

Magagula, G T, Ndlovu, L S, Sangweni, R and Dlamini, S (1996) Strategies and Policy Framework for Water Resources Management in Swaziland. Paper prepared for the Workshop on Economics, Policy and Natural Resources Management CSIR, South Africa.

Opschoor, J B and Vos, H (1989) *Economic Instruments for Environmental Protection.* OECD, Paris.

Simpson, L D (1994) Are Water 'Markets' a Viable Option? *Finance and Development* June 1994: 30–32, The World Bank, Washington, DC.

SMEC, WLPG, SGAB (1991) The Botswana National Water Master Plan (vls 1–12). Department of Water Affairs, Gaborone.

United Nations (1992) Report of the UN Conference on Environment and Development. Rio de Janeiro.

Water Utilities Corporation, several years' Annual Reports, Gaborone.

Zimconsult (1996) Water Pricing Options and Implications in Zimbabwe: Case Studies Paper (draft).

4

WILDLIFE USER RIGHTS IN ZIMBABWE

J W Arntzen[1]

BACKGROUND

The primary environmental issue addressed in this case study is the need to halt the decline of wildlife resources in communal areas in Zimbabwe, to create benefits for the local communities. Wildlife populations in communal areas have been declining for two reasons: (1) rapid population growth, requiring more land for agriculture and livestock; and (2) the perception of local people who see wildlife not as a resource but as a nuisance, since they suffer net costs from the presence of wildlife.

As National Parks are not sufficient to protect the current numbers of wildlife resources and as non-agricultural development opportunities are very limited, community based wildlife projects have been conceived to meet two important goals: promoting rural development opportunities and conserving wildlife resources in communal areas. Whilst this approach has been adopted by virtually all countries in the region, Zimbabwe has so far gained the most experience and has been most successful, primarily through its Communal Area Management Programme of Indigenous Resources (or CAMPFIRE). Neighbouring countries are in the process of copying or developing their own version of CAMPFIRE.

In recent years, macroeconomic constraints and related structural adjustment measures have limited the government's ability to provide and maintain basic services, including natural resource management and the promotion of rural development. The objectives of the CAMPFIRE programme are to:

- conserve the fragile environment and sustain economic viability through wildlife utilization;

- reduce the conflicts between wildlife and agricultural development by fencing;
- increase rural employment and incomes;
- create local institutions for indigenous resource management.

The underlying assumptions of the programme are that:

- rural people consider wildlife as a nuisance and not as a resource;
- wildlife cannot be conserved only in protected areas;
- wildlife utilization may be most suitable in marginal or degraded areas (zone 4 and 5).[2]

Generally, the more marginal the physical conditions become, the more attractive wildlife utilization becomes. However, it is important to recognize that in communal areas, wildlife and livestock cannot be considered substitutes. Livestock generates many products (including draughtpower and manure) not provided by wildlife. They are comparable only in terms of income generation.

DECENTRALIZATION OF USER RIGHTS

The main economic instrument used in the CAMPFIRE programme is the decentralization of wildlife user rights. With respect to the revenues generated, the following secondary economic instruments are used:

- auctioning of user rights;
- net revenue-sharing formula to ensure funds for resource maintenance and to cover administrative and monitoring costs;
- flexible, locally determined distribution of local revenues;
- compensation for wildlife damage.

The implementation of these instruments was expected to lead to a reversal in the trend of wildlife decline in communal areas and an increase in wildlife-related benefits for the local population.

In the past, wildlife conservation relied heavily on regulations and, because of enforcement problems, there was a vacuum in resource management. Under the new approach, the government has legislated that wildlife user rights be decentralized to councils and communities in CAMPFIRE areas (Department of National Parks and Wildlife Management, 1992). District Councils wishing to manage wildlife resources may be granted this authority subject to a number of conditions (including the development of a management plan and an institutional plan outlining decentralization to the communities, and the approval of all quotas by the Wildlife Department).

Central government delegates wildlife user rights to district councils, which in turn delegate these rights to local communities (upon their application). Based on annual wildlife counts, user quotas are set by the Dept. of National Parks and Wildlife Management for each community involved in the programme. With assistance from the government and some NGOs, communities prepare annual plans to utilize the wildlife resource. Basic uses are hunting and/or game viewing and additional revenue may be obtained through wildlife processing activities. Communities may either manage the wildlife resources themselves (with council assistance) or, as is more commonly the case, lease out the rights to the private sector (if, for example, communities need to bring in additional skills). There is a revenue-sharing formula ensuring that sufficient funds are available for resource management (35 per cent) and the local communities (at least 50 per cent). CAMPFIRE has proven popular and now involves more than 84 communities with some 400,000 inhabitants, covering an area of 30,000 km² (Bond, 1994).

IMPLEMENTATION ISSUES

The high overhead costs of CAMPFIRE projects have been met with the assistance of donors. Since Zimbabwe is in the second phase of an economic restructuring programme, councils have discovered that wildlife can be an important source of revenue. Wildlife revenues are highest in communities with abundant wildlife resources, in particular elephants and buffalo. Therefore, the project appears feasible only in areas with sufficient wildlife. Other areas would require expensive re-stocking measures. Project revenues have been adversely affected by the CITES ban on ivory trade, showing a drop in income of about 20 per cent (see Bond, 1994). The programme was intended to initially cover wildlife, and then to extend to other local natural resources. This extension has not yet happened. There are limited quantitative data on the environmental impacts of the programme, though research reports indicate that wildlife resources have stabilized in CAMPFIRE areas.

IMPACTS

The programme has been successful in raising revenue, with 90 per cent of the revenue coming from hunting (see Table 4.1). Elephants and buffalo make up 82 per cent of the hunting revenues (Bond, 1994). The revenue-sharing formula appears to have good compliance rates. Wildlife management received a lower percentage (22 per cent) than the target of 35 per cent, but the difference will presumably come from the as yet unallocated funds. Compensation to people for crop and animal losses by wildlife is paid from the allocation to the local population.

Table 4.1 *Sources and Allocation of Wildlife Revenue under Zimbabwe's Campfire Programme (1989–1992)*

Category	Absolute revenues (thousands Z$)[a]	% of total income
Total Income:	11,509	100
Hunting	10,307	90
Tourism	164	1
Problem animals' hides and ivory[b]	244	2
Other revenues	740	7
Allocation:		
District Councils	1339	12
Wildlife Management	2533	22
Local Population	5460	47
Other	298	3
Unallocated	1772	15

Notes: a Z$ = Zimbabwean dollars; Z$9.65 = US$1. b Revenue from the sale of hides and ivory of buffalo and elephant which had been attacking livestock or humans or causing other problems.
Source: Bond, 1994

The auctioning or tendering of wildlife user rights has more than doubled revenue over the period 1989 to 1993 (Child and Bond, 1994). Table 4.2 shows the trends in marketing hunting quotas from communal areas. While hunting quotas increased by 43 per cent between 1990 and 1993, income increased six times to an estimated Z$10.4 million in 1993. The costs of the ivory ban over the period 1989–1992 are estimated to be Z$4 million, or roughly 20 per cent of the generated revenue (Bond, 1994). Comparing CAMPFIRE progress in Kanyurira in the Zambezi valley between 1988 and 1993, Matzke and Nabane (1995) conclude that the development benefits are increasing in terms of income, security and local empowerment. The ability of communities to agree on the distribution of the local revenue enhances household security (for example, during droughts a large part of the revenue is given to households) and local development (when the need arises, a larger part of the money may be used for public services).

Table 4.2 *Trends in Efficiency of Marketing of Hunting Quotas to Safari Outfitters*

	1990	1991	1992	1993[a]
Value of quota (Z$)[b]	2,464,445	2,555,045	3,164,495	3,526,055
Income (Z$)	1,505,956	2,725,868	6,016,738	10,365,485
Income (US$)	602,382	1,090,347	1,203,348	1,645,315
Efficiency indicators				
Z$/standard value	0.61	1.07	1.9	2.94
US$/standard value	0.24	0.43	0.38	0.47
Z$/US$	2.5	2.5	5.0	5.3

Notes: a 1993 revenues are 'expected'. b Value of quota is estimated using standard prices.
Source: Child and Bond, 1994

A detailed comparison of the different marketing systems of user rights is given in Table 4.3. Substantial differences in income generation exist. Compared to the value of the quota (column 3), the real revenues are in a few cases slightly lower, but in most cases substantially higher. This is expressed in the last column. Tendering – and in particular auctioning – the quotas yields higher returns than set fees and allocations.

Limited data are available on income generation at the household level. The current agreement appears to be that CAMPFIRE provides valuable income supplements for households. Income data seem to range from as low as Z$24.25 per household (Nyaminyami) to up to Z$400 per household. Data provided by the WWF multispecies project in Harare (pers. comm. Bond) show that revenues are relatively modest for the majority of wards. In half of the wards, the average extra household income for the years 1989–1993 was Z$30 per household, with the top 10 per cent of the wards showing at least Z$180 more income per household. Interestingly, revenues were negatively correlated to population density. In other words, the higher the population density, the lower the wildlife revenues obtained. Murombedzi (1991) found that in the Nyaminyami CAMPFIRE project, incomes from wildlife were supplements to wage and agricultural income. The additional revenue was mostly invested in crop production. The optimal distribution of local revenue appears to be a mix between community projects and individual household income allowances. In drought years, income allowances are increased to compensate for agricultural income losses and to contribute to livelihood security. The project appears to have gained wide acceptance, as evidenced by the increased participation of local communities. CAMPFIRE is gradually changing people's perception of wildlife, to see wildlife as a resource. CAMPFIRE may increase Zimbabwe's competitiveness for tourism as communal areas remain attractive. Finally, the benefits of CAMPFIRE may spread in time to other natural resources or to development projects because of the successful establishment of local management institutions.

Conclusions

The conclusions which can be drawn are rather tentative since, although there are many reports on CAMPFIRE, few give details of any evaluations of the programme.

The instrument is an innovative combination of two key elements in resource management:

- increased use of economic instruments to generate benefits;
- decentralization of resource management and revenue generation.

Table 4.3 *A Comparison of Wildlife User Rights Marketing Systems (US$; 1992)*

Area	Marketing Channel	Standard Value of Quota (US$)	Expected Income	Received Income	Average Income/ Standard Value
Communal areas					
	Renegotiated	995,450	1,527,000 (5 areas)	1,814,073 (5 areas)	2.29 (6 areas)
	Tender	1,322,635 (11 areas)	3,942,045 (10 areas)	4,074,055 (11 areas)	2.52 (11 areas)
	Interview	73,600	300,000	230,000	3.13
Parks, wildlife estates					
	Tender	400,250 (2 areas)	–	449,972 (1 area)	1.85
	Auction	1,268,055 (2 areas)	–	5,834,166 (2 areas)	5.86
	Set fees	1,800,475 (3 areas)	–	3,041,685 (3 areas)	1.41
	Lottery	241,200 (1 area)	–	178,580 (1 area)	0.74
	Allocation	114,725 (1 area)	–	121,840 (1 area)	1.06

Source: condensed from Child and Bond, 1994.

Annual wildlife counts provide the basis for the conservatively set hunting quotas. These quotas appear to address a genuine resource issue, as testified by their popularity. However, some concerns remain regarding the sustainability of the programme. First, the revenues are mostly from hunting, in particular of elephants and buffalo; over-hunting can easily happen. Therefore, quotas must be carefully set (as they presently appear to be). However, pressure for higher quotas is likely to increase if economic hardships persist or wildlife resources decline. Second, the future of the wildlife resource in communal areas is determined by several factors, many of which are beyond the control of the local communities. Elephants and buffalo require large areas in which to roam, and population growth and resettlement may cause reduction and fragmentation of wildlife areas, leading to the disappearance of these key species. Third, programme revenues are an important and growing income source for district councils. To avoid councils appropriating large portions of the revenue, the existing formula for revenue sharing should be strictly applied.

The present institutional structure appears to function well, but there may be pressures (related, for example, to structural adjustment) to modify the role and revenue of each institution involved. It is unclear how permanent the present structure actually is.

NOTES

1   University of Botswana.
2   The term wildlife utilization refers here to the managed exploitation of wildlife resources, including the hunting of game and the development of safari viewing for tourists.

REFERENCES

Barnes, J I (1988) The Economics of Wildlife Utilisation Options. In: KCS (1988) *Sustainable Wildlife Utilisation: the Role of Wildlife Management Areas.*

Barnes, J I (1994) Alternative Uses of Natural Resources. In *Botswana Society 1994 Botswana in the 21st Century.* Gaborone.

Bond, I (1994) The Importance of Sport-Hunted African Elephants to CAMPFIRE in Botswana. *Traffic* vol 14 (3): 117–119.

Child, B and Bond, I (1994) Marketing Hunting and Photographic Concessions in Communal Areas. In: Jones, M A (ed), Proceedings of the Natural Resources Management Seminar: Safari Operations in Communal Areas in Matabeleland. Department of National Parks and Wildlife Management, Gaborone.

Conybeare, A and Rozemeijer, N (1991) *Game Ranching in Botswana.* Government of Botswana/USAID.

Child, B and Peterson, J H (1991) *Campfire in Rural Development: the Beitbridge Experience.* CASS/University of Zimbabwe and Department of National Parks, Gaborone, Botswana.

Department of National Parks and Wildlife Management (1992) *Policy for Wildlife.* Ministry of Environment and Tourism, Harare.

Hawkes, R K (1991) *Crop and Livestock Losses to Wild Animals in the Bulilimamangwe Natural Resources Management Project Area.* CASS, University of Zimbabwe.

Martin, R B (1993) *Should Wildlife Pay its Way?* Department of National Parks, Government of Botswana.

Martin, R B (1994) *The Influence of Governance on Conservation and Wildlife Utilisation.* Department of National Parks and Wildlife Management, Zimbabwe.

Matzke, G E and Nabane, N (1995) African Wildlife Conservation, Utilisation and Community Empowerment: Zambezi Developments Continue. *Ambio* vol 24 (5): 318–319.

Muir, K (1993) *Economic Policy, Wildlife and Cattle Management in Zimbabwe and their Environmental Implications.* AFTEN, The World Bank, Washington, DC.

Murombedzi, J (1991) Decentralising Common Property Resources Management: A Case Study of the Nyaminyami District Council of

Zimbabwe's Wildlife Management Programme. Dryland Networks Programme Paper no 30, IIED, London.

Murphree, M W (1991) *Communities as Institutions for Resource Management.* CASS-University of Zimbabwe.

Murray, M L (1978) *Wildlife Utilisation Investigation and Planning in Western Botswana.* Ministry of Commerce and Industry, Gaborone.

5

HUMAN AND SOLID WASTE MANAGEMENT IN URBAN AFRICA: A CASE STUDY OF ACCRA, GABORONE AND HARARE

J W Arntzen and N H Fidzani[1]

BACKGROUND

Rapid urbanization has led to urban congestion and the rapid growth of slum areas in most African cities. This has naturally placed a burden on the provision and maintenance of adequate sewage and solid waste facilities, particularly in countries with stagnating economies. Human waste is the most important cause of water pollution, and a serious problem in many low-income countries (World Bank, 1992). Proper treatment of human and solid waste constitutes a key element in maintaining the quality of the urban environment. However, waste treatment is frequently subject to market failures, making corrective or preventive economic instruments necessary for four reasons (Porter et al, 1995). First, the consumption waste treatment at appropriate levels creates positive benefits to other people. Second, waste treatment is to a large extent a public good because of non-rivalry and non-exclusivity in consumption. Third, waste treatment is a merit good in that the society expects each person to have access to a certain minimum consumption level. Finally, the high infrastructural requirements imply that the fixed costs are high, which discourages participation by many suppliers. The resulting monopolistic tendency may require government intervention.

It was difficult to find a case study of environmental economic instruments used for pollution control for Africa. In South Africa the use of such instruments is being considered and feasibility studies have been carried out (see eg Economic Project Evaluation, 1993a and b). This chapter is based primarily on a comparative study of human and solid waste disposal in three African cities by Porter et al (1995). The reliance on limited literature may affect the scope and detail of the case study. However, the comparison between three African capitals offers valuable insights into the operation of economic instruments under different economic and social conditions. Moreover, the urban focus provides a useful addition to the other, rural, African case studies.

ENVIRONMENTAL ISSUES

The environmental issue addressed in this study is an effective and efficient disposal system for human and solid waste, to reduce the wastes' adverse impacts on the environment (such as sanitation problems). The assumption is that the removal of both types of waste from residential areas is more important than the application of advanced waste treatment processes (such as landfill or incineration). The primary methods of human waste collection are sewerage systems, septic tanks, and pit or pan latrines. With respect to solid waste, the collection systems include: location of containers (in commercial and low income areas), regular bin collection and special arrangements for 'odd' waste (eg garden waste or building rubble).

ECONOMIC AND SOCIAL CONTEXT

The comparison of three countries yields some interesting insights into the impact of the socio-economic context on environmental management. Some of the key characteristics have been summarised in Table 5.1. There is a striking difference in the economic performance of, on the one hand, Botswana (with a rapid growth, and a relatively high per capita income) and on the other hand Ghana and Zimbabwe (with negative growth and declining per capita income). In essence, Accra, the capital of Ghana, with a population of around 1.1 million in 1992, is characterized by relatively slow population growth (3 per cent per annum), emigration of skilled labour and limited funds for environmental management. Crowding has led to inadequate and deteriorating infrastructure; Accra has an open drainage system. Botswana is characterized by rapid economic growth and urbanization. Economic growth was high in the period 1970–1990, although it has declined to between 3 and 5 per cent in the early 1990s; Gaborone, the capital of Botswana, has been growing very fast but still remains a small city with an estimated population size of around 150,000 in 1996. Despite abundant government revenues, the policy of restricting urban subsidies has led to some squatter settlements in Gaborone. Zimbabwe's waste treatment system has been shaped by its history and the present economic conditions in the country, which has experienced severe economic problems in recent years. The post-independence government inherited a relatively good (and expensive) waste disposal system, which despite racial segregation provided sewerage systems to most residential areas (high and low-income groups). Ironically, the high costs of the system have now become a burden for the government and the population. Racial segregation rules suppressed urbanization for a long time, but in recent years urbanization has increased.

Despite the substantial differences in per capita income, per capita solid waste generation is the same in each country (around 0.7 kg/day, of which two thirds is residential; Porter et al, 1995).

Table 5.1 *Socio-Economic Characteristics by Country*

Characteristic	Ghana	Zimbabwe	Botswana
Population (million; 1992)	15.8	10.4	1.4
Population density (people/km^2; 1992)	66	27	2
Urban population (as per cent of total; 1992)	35	30	27
Percentage of population living in the capital (1990)	4.3	5.9	8.8
GNP per capita (US$; 1992)	450	570	2790
Annual per capita income growth (1980–92)	–0.1	–0.9	6.1
Net present value of external debts as per cent of exports (1992)	236	187.2	n.a.
Net present value of external debts as per cent of GNP (1992)	39.1	63.8	12.6

Source: World Development Report 1994

USER CHARGES FOR WASTE DISPOSAL AND TREATMENT

A variety of user charges (Panayotou, 1994) are being (or have been) used in the three cities (see Tables 5.2 and 5.3 in the Annex). Solid waste from households is collected once or twice a week, and the following charges are (or have been) used: monthly solid waste charges, pay-as-you dump fees or a collection charge as part of a monthly service fee (for low-income groups in Gaborone) or annual municipal rates (for house owners in Gaborone and Harare). For commercial waste, public containers are used and collected at a charge. With respect to human waste, various systems are in use: sewerage systems, septic tanks, pit and pan latrines and public toilets. Economic instruments include a connection fee for sewage systems and a monthly charge (in Accra).

The expected results are not explicitly indicated, but it appears that two results are expected: a better urban environment and greater cost recovery, at least breaking even with the operational costs. Waste collection and disposal systems are expected to lead to healthier and more pleasant urban living conditions for the entire urban population. Moreover, proper waste treatment reduces the risks of environmental damage from uncontrolled waste disposal. No explicit reference is made to waste reduction or recycling objectives.

OTHER INSTRUMENTS

These economic instruments are being used in conjunction with the following associated instruments:

- *legislative instruments* (including ordinances and a ban on pit latrines);
- *institutional reforms:* for example, in Accra the management of public toilets was successfully decentralized to the local level. In Accra and Gaborone, solid waste collection was partly privatized. The results were positive, and the collection became cheaper. Finally, the first stage of human waste collection from pan latrines (ie bringing it to collection depots, from where the waste management department collected containers at US$15/container) was privatized in Accra;
- *subsidies:* most charges are only meant to cover operational costs. Capital costs are therefore implicitly subsidized. Two forms of subsidy occur: cross-subsidization from high to low income groups,[2] and government subsidies.

The implementation of some instruments was discontinued as they failed to work as expected. Examples include: collection fees for solid waste in Accra, and the pay-as-you dump charge in Accra.

DESIGN AND IMPLEMENTATION ISSUES

Three design problems became apparent. Firstly, it proved difficult to identify which collection system would be most suitable for the socio-economic conditions. The collection system must take into account people's incomes and willingness to pay. For example, sewerage systems may be considered too expensive by people, and connections to the system may increase only if they are made compulsory (as in Accra). The introduction of sewerage systems to low income areas in Gaborone may prove to be very costly to both the government and the users (who face a waste charge and a higher water bill). In addition, water consumption may be too low to avoid sewage blockage. The risk of free riding is also high, especially when the waste disposal is entirely a public good (see below). A second design problem was how to cover costs, in particular capital costs but even operational costs. Most schemes suffer from maintenance problems and the necessary recurrent expenditures are not

properly planned or can no longer be afforded. Systems with low recurrent costs for government therefore have a clear advantage (such as
systems which are privatized). A third problem was how to link the level
of charges to the amount of waste generated. As they stand, the charges
are not likely to reduce the amount of waste; their primary contribution
lies in financing the collection schemes.

In the design of these instruments, the nature of the good needs to
be analysed. In this respect, there is a small but essential difference
between human waste and solid waste disposal (Porter et al, 1995). The
former has an element of private good (privacy) whilst the latter is
entirely public. This would make it difficult to impose substantial charges
for the latter. Environmental economic instruments need to be designed
in conjunction with other regulatory and consultative instruments. For
example, high income groups in Harare still have the option to have
septic tanks in their yards, instead of connecting to the more expensive
sewerage system.

These problems appear to revolve around maintenance, participation
and revenue collection. The administrative costs of modest fees are very
high, and governments may find it difficult to implement an efficient
revenue collection system. If left a choice, households may opt not to
link up with sewerage systems, so economic instruments need to be
backed up by legal requirements. Moreover, the cost recovery issue
should be addressed in the design stage (either through cross-subsidization or government subsidies).

ENVIRONMENTAL IMPACTS

The following data on waste recovery rates are available:

Solid waste: Harare: 90–95 per cent; Accra: less than 50 per cent and
declining.
Human waste: Harare: very high through sewerage system (with proper,
expensive treatment);
Accra: not precisely known but probably also declining;
Gaborone: no recovery data were available.

In general, the environmental results have been mixed. In Accra, the
amount of solid waste collected has declined to between 40 and 50 per
cent. There is also a downward trend in human waste collection (90 per
cent was collected in 1992). Most instruments for solid waste disposal
have been discontinued because of implementation failures (including
collection fees, monthly fees, and pay-as-you dump fees). Decentralized
management has also had limited success in terms of cost recovery. The
most promising measure has been to privatize solid waste collection in
middle income areas. This has actually led to lower collection fees

(US$2.50 compared to US$4.40 charged by government). Linking charges with regulations helped to increase the number of sewerage connections in Accra. A constant constraint was the danger of the free-rider problem (especially with solid waste).

It should be noted that the study does not really address the environmental impacts of the discharge (after or without treatment) of the collected waste. Such environmental impacts are determined by local environmental and socio-economic conditions. Ghana may have an advantage with its higher rainfall that can substantially dilute any water contamination.[3] In this respect, there is a clear contrast between Accra and Harare. In Accra, the human waste collected from pan toilets and the sewerage system is subsequently discharged without any treatment (nightsoils are dumped on the beach and washed into the sea). Sludge from septic tanks is used as fertilizer (and sold at US$60/truckload against a tipping fee of US$5/truckload; Porter et al, 1995). Obviously, high levels of discharge may have substantial environmental impacts. By contrast, in Harare, the sewerage system is much more widespread and receives extensive and expensive treatment. The resulting water is used for irrigation. Solid waste is brought to a huge landfill (without treatment; a tipping fee of US$5 to US$10 is charged, depending on the size of the vehicle). Solid waste is not treated in landfills, which has led to frequent fires.

ECONOMIC AND SOCIAL IMPACTS

Most economic instruments aimed at achieving cost recovery of the variable costs involved. However, participation and/or revenue collection problems frequently made this impossible. People's willingness to pay was probably adversely affected by the poor (and deteriorating) quality of services. This posed particular problems in Accra and Harare where government subsidies had to be reduced. Privatization may be a way out, as the private sector has proven more innovative in the use of low cost, labour-intensive collection methods, and at least has to break even. Nonetheless, it will still be important to maintain a degree of government supervision.

It appears virtually impossible to achieve cost recovery for solid waste disposal in low income areas. This problem can be solved through subsidization by other waste generators or from general government revenues.

Waste collection methods and their success are affected by factors such as the housing market (government-dominated or mostly private), the history of the cities (for example, compared with Accra, Gaborone is a relatively new city and well serviced with a sewerage system).

CONCLUSIONS

The choice of economic instrument should be considered in close association with the choice and design of the collection systems. If this is not done, there may be many implementation problems, and cost recovery may be even more difficult than is normally the case.

The nature of the good also influences the effectiveness of the economic instrument. Whilst human waste charges may be acceptable because its collection is not entirely a public good, it is very difficult to charge specifically for solid waste collection. High charges encourage illegal dumping. Expenditures could be met from general government funds or taxes. However, cross-subsidization from high to low-income groups could be justified by the higher willingness (and ability) to pay.

Charges in relation to the amount of waste generated were rare and appeared successful only in the public toilets of Accra. Charges for special solid waste consignments of household solid waste appear to work fairly well (a certain amount per truck). The pay-as-you dump charge for solid waste resulted in unacceptable free-rider behaviour.

Institutional flexibility is important. A monopoly supply of services by central government proved to be unsuccessful, particularly under difficult macro-economic conditions. Decentralization to the local government level could improve the situation. In the case study, privatization of solid waste collection proved to be most successful, from both an environmental and economic viewpoint.

Legal enforcement may be necessary to break through the cycle of low participation (in sewerage connection) and high costs per connection, which can make it less attractive to join and can result in government deficits.

Revenue collection proved to be a major obstacle. A solution attempted in Gaborone was to lump small service fees together into a monthly service charge for water and human/solid waste collection or an annual charge for house owners in other areas.

The case studies also revealed substantial problems in the delivery of quality services, caused mainly by the difficult economic conditions. Even in Gaborone, where economic conditions were somewhat better, the quality of service (for example, the breakdown rate of the waste collection trucks) could be improved.

REFERENCES

Economic Project Evaluation (1993a) The Use of Economic Instruments to Control the Discharge of Chloride Rich Effluent to the Crocodile River. Research Report no. 4, Department of Environmental Affairs, South Africa.

Economic Project Evaluation (1993b) The Use of Economic Instruments to Control the Discharge of Effluent from Sappi's Ngodwana Paper Mill to

Elands river. Research Report no. 3, Department of Environmental Affairs, South Africa.

Panayotou, T (1994) Economic Instruments for Environmental Management and Sustainable Development. Report prepared for UNEP. Harvard Institute for International Development, Massachusetts.

Porter, R C, Boakye-Yiadom, I, Mafusire, A and Tsheko, B O (1995) The Economics of Water and Waste in Three African Capitals. University of Michigan, mimeo.

World Bank (1992) *World Development Report 1992*. Oxford University Press.

ANNEX: SUMMARY OF ECONOMIC INSTRUMENTS USED FOR HUMAN AND SOLID WASTE COLLECTION

Table 5.2 *Summary of Economic Instruments for Human Waste Collection*

City	Instrument	Operation	Constraints	Strengths
Accra	sewage connection charge + 25–35% surcharge on the water bill	1978–present	needs ban on septic tanks in central Accra	
	septic tank cleaning fee aimed at breaking even (same for government and private)	present	collection problems	
	pan latrine collection fee ($2–3/month);	up to 1987		
	$0.02 charge per use of public toilets; revenues used to keep toilet clean		payment problems by local councils; too few public toilets; fees act as deterrent	involvement of local councils has led to cleaner toilets
Harare	most low and high income groups are connected to sewage system, and pay annual fees. Fees for poor reduced through cross-subsidization and government subsidies (coverage of running costs only)	present	fees too high for low income groups; difficult for government to subsidize	
Gaborone	pit latrine fees are part of monthly service charge to low income areas; sewage charges are part of an annual charge for property owners (proportional to the property value).		collection problem	

Source: Porter et al, 1995

Table 5.3 *Summary of Economic Instruments for Solid Waste Collection*

City	Instrument	Operation	Constraints	Strengths
Accra	2$ refuse levy fee as surcharge on property tax; pay-as-you dump fee ($0.02/household container)	1983–1993	collection problems (10% full compliance rate) payment failure; system never covered more than 30% of the running costs; collection difficulties	
	decentralization of charge collection to local authorities	1993–94		
	for mixed and low-income areas: privatization	1995		
Harare	refuse fees: initially as an addition to 6-monthly property tax. Now: a separate monthly charge. In effect, average households paid $2 per annum on solid waste fees in the past decade		only covers operational expenditures	
Gaborone	solid waste fees are part of monthly service charge in low-income areas;	present	collection problems	
	solid waste charges are part of an annual charge for property owners (proportional to the property value).	present	collection problems	

Source: Porter et al, 1995

NOTES

1   University of Botswana.

2   Interestingly, if charges were related to the quantity of waste generated, low income groups would be found to be subsidizing the high income groups because of the higher volume of waste generated by the latter.

3   Emission in low-flow or dry rivers may cause considerably poorer water quality as is demonstrated by the case of the Eland and Crocodile rivers in South Africa (Economic Project Evaluation, 1993a and b).

PART II

ASIA CASE STUDIES

6

MALAYSIAN PALM OIL POLLUTION TAX

Adis Israngkura[1]

BACKGROUND

During the period 1970–1985, Malaysia's palm oil industry experienced a rapid expansion. Production levels reached about 4.2 million tonnes in 1985, with palm oil exports accounting for about 10 per cent of Malaysia's total exports. This volume of export represented as much as 75 per cent of the world trade in palm oil in 1980. Alongside this expansion, palm oil production also created water pollution problems. Measured in terms of biochemical oxygen demand (BOD), in 1978 the level of pollution from palm oil was more than 1.5 million tonnes of BOD per day – equivalent to the amount of pollution created by a population of 15 million, roughly the entire population of Malaysia at that time. This situation created a dilemma for the Malaysian Government as the palm oil industry was the country's major income earner yet was creating an environmental problem which the country could not afford.

Palm oil extraction mills discharged their effluent containing soluble organic compounds into rivers. These dissolved organic compounds are decomposed naturally by aerobic (oxygen-requiring) bacteria. However, when released in large amounts, the level of oxygen in water can be depleted, affecting the river ecosystems. It was reported that by the late 1970s, many rivers in Malaysia were heavily polluted and the fish population, which provided food to the rural poor, had been devastated. Palm oil-related pollution also affected the country's mangroves (which serve as nurseries for many aquatic animals) and hence had a negative impact on marine and brackish-water fish populations as well. Besides its effect on aquatic animals, water pollution also meant that river water could no longer be used for consumption purposes and many rural families had to switch to other sources of water supply such as groundwater.

At this stage it seemed inevitable that action was needed from the government to end the pollution generated from an economic activity important to the national economy.

ECONOMIC INSTRUMENTS FOR POLLUTION CONTROL

The aim of the Malaysian Government, therefore, was to reduce palm oil-related water pollution while at the same time maintaining the expansion of the palm oil industry. The Malaysian Government aimed to reduce the average effluent discharge from 220 tonnes per day to 25 tonnes per day.

In 1977 the Environmental Quality Regulations specified standards on eight variables concerning palm oil effluent discharge: biochemical oxygen demand (BOD), chemical oxygen demand (COD), total solids, suspended solids, oil and grease, ammoniac nitrogen, organic nitrogen, and the level of acidity measured as pH. These standards became more stringent through time. Table 6.1 shows the palm oil BOD standards specified during 1978–1984.

Table 6.1 *Effluent Standards for BOD, 3-Day at 30° Celsius (milligrams/litre)*

	Standard A	Standard B	Standard C	Standard D	Standard E	Standard F
Year	1978	1979	1980	1981	1982	1984
BOD_3	5000	2000	1000	500	250	100

Source: Department of Environment, Environmental Quarterly Report 1981–1984

To solve this water pollution problem the Malaysian Government employed a combination of instruments – a pollution tax that was referred to as a licence fee, effluent standards, and R&D subsidies in the form of tax exceptions. These instruments became effective in July 1978.

Mills which released pollution loads within the specified standards set by the Department of the Environment (DoE) were subject to a normal licence fee (regular pollution tax). If a standard was violated, the mills were subject to an excess charge equivalent to 10 times the licence fee.

By tightening the effluent standards each year, the DoE adopted a gradual approach to improving the quality of river water. The mills were informed beforehand that these effluent standards would become more stringent each year. This warning was intended to induce the mills to invest early in treatment facilities.

Malaysia also recognized the importance of R&D activities in providing pollution treatment technology to the palm oil industry. To achieve this objective, the DoE granted a total or partial waiver of the licence fee if the mills conducted in-house R&D programmes.

The palm oil pollution tax used in Malaysia was in the form of a licence fee. There were three kinds of licence fee:

1   an administrative fee of M$100;[2]
2   a fee of M$10 per tonne of BOD load discharged into watercourses or a fee of M$0.05 per tonne if discharged on land; and
3   in cases where BOD exceeded the standard in effect that year, an excess fee of M$100 per tonne of BOD load for both watercourse and land disposal.

These fees were intended to encourage mills to reduce their pollution in a least cost manner. Mills could choose either to dispose of their pollution and pay the licence fee or to invest in a treatment facility and save on the licence fee. It was believed that mills would make choices based on the level of their palm oil output (and hence the pollution level) that would determine which of the two choices was profit maximizing for them.

It is worth noting here that these licence fees were set arbitrarily at levels that were believed to be high enough to reduce palm oil pollution. The fees were not calculated from the marginal benefit from environmental improvement and hence might not maximize net social benefit. To maximise net social benefit from environmental improvement, the fees would have needed to be calculated based on the marginal benefit and the marginal cost of pollution reduction.

The revenue from the palm oil licence fees was used to promote R&D in palm oil effluent treatment. The revenue was not allocated directly to any public or private R&D programme, but rather was used to finance the waiver scheme whereby mills with in-house R&D activities were entitled to a partial or total exemption of the licence fees. This scheme can be considered as an implicit R&D subsidy.

In 1978, the year in which the licence fee was launched, the total revenue generated was around US$110,000. The revenue declined to around US$28,000 in 1979–80 and by the end of the 1980s it had stabilized at around US$11,000–US$14,000. It may be noted that the revenue collected in the first year was unusually high as many mills could not adjust to meeting the effluent standard and chose to pay the excess fee instead.

IMPLEMENTATION ISSUES

In implementing the regulation, the DoE required mills to file an application to obtain operating licences. In the application mills had to specify the type of technology they planned to employ to reduce palm oil pollution. Mills must regularly file reports to the DoE stating the level of pollution actually discharged into the watercourses. The DoE also made it clear to the mills that the standards used would become more stringent each year and within four years the mills must reduce their BOD from 5000 ppm (parts per million) to 500 ppm.

The Environmental Quality Act was adopted in 1974 and the DoE was established in 1975. To set the conditions for granting the licences the DoE formed a committee comprised of representatives from both the government and the private sector. The task of this committee was to ensure that the instruments used were appropriate environmentally, technologically and financially.

The DoE issued licences to the mills, the licence fee being set according to:

- the class of premises;
- the location of premises;
- the quantity of wastes discharged;
- the pollutants or class of pollutants discharged; and
- the existing level of pollution.

An economic assessment of the palm oil licence fee system rests mainly on how these fees are calculated. A licence fee scheme that can ensure economic efficiency in resource allocation is one where the fee is set according to the marginal damage of pollution. This implies that for the licence fee to be economically efficient it must induce mills to control pollution up to the point where the marginal benefit from pollution reduction equals the marginal cost of doing so. When the fee is calculated in this manner, firms will invest resources in environmental control efficiently and the society at large will benefit from pollution reduction.

For instance, when a mill finds it cheaper to pay the fee than to treat the pollution it implies that the benefit from treating the pollution (reflected in the amount of fee payable) is lower than the cost of pollution control. In this case, the mills will continue to pollute and pay the fee. On the other hand, if the benefit of treating pollution (ie the fee) is higher than the cost of controlling pollution, mills will choose to treat pollution instead of paying the fee. This will ensure that pollution control is carried out in a manner that would yield maximum net social benefit to the society.

However, the licence fee used to control Malaysian palm oil pollution was not calculated in this manner, since its aim was to induce mills to reduce their average daily pollution to a goal of 25 tonnes. This environmental objective could prove to be inefficient from the perspective of efficient resource allocation as it might induce too little or too much investment in pollution control.

On the other hand, the licence fee scheme was efficient in being able to control pollution at the lowest cost possible. It did this by encouraging lower cost mills (ie mills which treat pollution at lower costs) to treat more pollution, while encouraging high cost mills to treat less pollution and pay the licence fee instead.

In the long run, however, this licence fee scheme has proven inefficient for three reasons. Firstly, there was no attempt to re-evaluate and adjust the fee to reflect the marginal benefit of environmental improvement. Fixing the licence fee at M$10 per tonne of BOD discharged into a watercourse did not guarantee that maximum net social benefit was realised from pollution control.

Secondly, the magnitude of the licence fee was never adjusted for inflation, leading to a decline in the real value of the fee over time. The fees of M$10 and M$0.05 per tonne of BOD for discharges into a watercourse and on land respectively, were never revised after their implementation in 1978. In real terms the value of the licence fee declined to about M$6.37 for water disposal and M$0.03 for land disposal.

Thirdly, as the real value of the licence fee eroded over time and the standard used for the excess fee became more stringent, it could be said that the means for pollution control gradually switched from providing economic incentives (licence fees) towards a command and control approach, namely, effluent standards. This gradual transition from effective licence fees to mandatory effluent standards probably made pollution control costlier to Malaysia as it forced all mills to treat pollution equally regardless of their cost structures.

IMPACTS

In terms of industry expansion it was argued that the use of a licence fee and effluent standard did not have a negative impact on palm oil businesses (Vincent, 1993). After the licence fee had been implemented for two years, the number of mills had increased from 131 to 147 and the output of palm oil had risen from 1.8 million to 2.6 million tonnes. However, by itself this increasing trend does not accurately depict the effect of the licence fee on the palm oil industry.

A more reasonable analysis would be one that compares the *actual* expansion of the palm oil industry against what it *would have been* in the absence of the licence fee and effluent standards. It is possible that the palm oil industry would have expanded at a higher rate had the licence fee not been implemented. This discrepancy between the actual growth and the growth that would have been possible would indicate the true effect of the licence fee on palm oil output. Unfortunately, forecasts of the growth of the industry in the absence of the licence fee and standards were not carried out, and so no meaningful analysis can be made on the impact of the licence fee on industry output. One piece of information that may indicate the negative effect of the licence fee is the decline of export share. Although the absolute volume of export increased, the share of palm oil export earnings declined from around 12 per cent in 1984 to about 7 per cent in 1989. This information certainly indicates how the palm oil industry grew at a slower rate compared to other sectors of the economy. Part of this decline in growth rate could be attributed to the use of the licence fee and effluent standards.

The palm oil licence fee also created a distribution effect where the burden of environmental improvement fell on palm oil producers and not on palm oil consumers. As palm oil is sold at a competitive price in the world market, the increase in the cost of pollution control can not be passed on to the consumers. When the export demand for palm oil is price elastic, much of the burden of a cost increase will fall on producers, mostly palm oil producers. By 1984, it was calculated that mills had spent about M$100 million on treatment facilities by mills. This amount could represent the loss incurred by palm oil producers.

The environmental results can be summarized as follows. In the first year the DoE aimed to reduce the mills' average discharge of BOD from 220 to 25 tonnes per day but the actual BOD only decreased to 125 tonnes per day. In the second year when the DoE raised the standard used for excess fee from BOD concentration of 5000 ppm to 2000 ppm and made these standards mandatory, the mills were able to reduce their average BOD discharge to 60 tonnes per day. However, this reduction still fell short of the goal of 25 tonnes per day.

In addition to inducing an overall reduction in the average discharge, the implementation of the licence fee also prompted innovations in treatment technology. A pond system was found to be suitable for those mills with access to land, while a system of agitated tanks was used by mills with limited land. These technologies proved simple and cost-effective to the palm oil industry.

In the fourth year, when the maximum permissible level of BOD was decreased to 500 ppm, a survey revealed that as many as 90 per cent of the mills were complying with this standard, and about 40 per cent of the mills had daily BOD discharges of less than 100 ppm.

In addition to the reduction in average pollution there was also an innovation in the use of palm oil by-products. During the early 1980s palm oil by-products were converted into animal feed and palm oil pollution on land was converted into fertiliser. Mills using tank digesters were also producing methane, which was used for generating electricity.

CONCLUSION

Malaysia can be considered the first country in South East Asia to seriously adopt an economic instrument to control pollution. The implementation of a licence fee combined with the use of effluent standards was the main feature of Malaysian palm oil pollution control. This economic instrument succeeded in reducing considerably the level of pollution, but still fell short of the target that had been set.

Unfortunately, implementation of the licence fee can not be accurately evaluated in terms of its economic impact. The possibility that the licence fee may have had a negative impact on the palm oil industry can only be guessed from the industry's declining export share over time. Furthermore, as the calculation of the licence fee was not based on the marginal benefit of environmental improvement, it would not ensure maximum net social benefit. As palm oil is sold at a competitive price in the world market, the burden of pollution control will largely fall on the Malaysian palm oil producers and not on consumers.

NOTES

1   National Institute of Development Administration, Thailand.
2   M$ = Malaysian Ringgit; approximately M$2.5 = US$1.

REFERENCES

Vincent, J R (1993) Reducing Effluent While Raising Affluence: Water Pollution Abatement in Malaysia. Working paper, Harvard Institute for International Development, Massachusetts.

Organisation for Economic Co-operation and Development (OECD) (1994) Applying Economic Instruments to Environmental Policies in OECD and Dynamic Non-Member Economies. OECD Documents, Paris.

Panayotou, T (1994) Economic Instruments for Environmental Management and Sustainable Development. Report submitted to UNEP. Harvard Institute for International Development, Massachusetts.

7

THAILAND'S UNLEADED GASOLINE PRICE DIFFERENTIAL

Adis Israngkura[1]

BACKGROUND

The current levels of lead in the environment are the result of economic activities, particularly the heavy use of leaded gasoline in automobiles. Emissions from automobiles using leaded gasoline have been a major contributor to the high levels of lead found in major cities with heavy traffic, such as Bangkok. In such traffic-congested areas, lead can be found in the air, water or food. When lead enters the human body and accumulates at high levels it can produce severe health effects, including damage to the central nervous system and chronic toxicity (Suwanna and Chamaiphan, 1994).

Lead released from the exhausts of automobiles using leaded gasoline, as well as lead from factory emissions, accumulates first in the atmosphere and then settles on water, soil and other substrata. Food sold on the side of busy roads is thus exposed to high levels of lead concentrations. In this way, lead can enter the human body through the respiratory system, dietary route or pores of the skin. Once it enters the body it can accumulate and spread to various parts of the body, such as the brain, lungs, liver, spleen and bones.

At high levels, lead in the bone marrow inhibits the production of haemoglobin. It can also damage the central nervous system, cause inflammation of the blood vessels and cause the brain to become swollen. Lead can cause abnormal behaviour in children, making them irritable or mentally retarded. High levels of lead in the blood also affect children's learning abilities. In such children, intellectual development tends to be slower: they have problems reading, interpreting words or concentrating. In adults, long term accumulation of lead can result in pain, dizziness, unconsciousness, fatigue and a feeling of disorientation.

Lead can also damage kidneys and the heart. Chronic exposure to lead causes deformities in kidneys and swelling of heart muscles. At high levels, lead can also affect human reproductive organs, causing sterility in males and premature births in pregnant women.

Many governments have attempted to solve the problem of lead pollution by encouraging road users to switch to using unleaded gasoline. Unleaded gasoline was first introduced into Thailand in 1992. In an attempt to alter consumer behaviour away from conventional leaded gasoline towards unleaded gasoline, the Thai Government adopted a pricing scheme known as a 'price differential'. Since the consumption of leaded gasoline creates environmental problems, the price differential aims to set the price of leaded gasoline at a level higher than the price of unleaded gasoline, to induce more consumers to use unleaded gasoline.

The Gasoline Tax Differential

The gasoline tax differential attempts to widen the price difference between leaded and unleaded gasoline by making leaded gasoline relatively more expensive, to encourage people to switch from leaded to unleaded gasoline. As the consumption of leaded gasoline drops, one would expect a corresponding reduction in the lead content in the air.

In addition, the government adopted a command and control approach, by requiring automobiles of 1600 cc capacity or higher, which had been assembled after 1 January, 1993, to have catalytic converters installed. From 1 September, 1993 onwards, only automobiles with catalytic converters were allowed to be registered.

The aims of establishing the gasoline tax differential were to:

- help reduce the level of lead in the air, thereby helping to reduce the associated health hazards and the impacts on children's mental development; and
- help reduce other kinds of air pollution, namely carbon monoxide, hydrocarbons and nitrogen oxide emissions, and the health risks associated with these gases, by requiring the installation of catalytic converters.

Implementation Issues

The following steps were taken by the Thai Government to encourage the use of unleaded gasoline.

Phase I
1   Introduction of unleaded gasoline retail sale on 1 May, 1991.

2   Reduction of the lead content in regular gasoline from 0.4 gm/l to 0.15 gm/l, starting 1 January, 1992.
3   Requirement that automobiles with capacities of 1600 cc and higher, assembled after 1 January, 1993, must have a catalytic converter and from 1 September, 1993, onwards, the requirement that only automobiles with a catalytic converter could be registered.
4   Promoting the use of unleaded gasoline throughout the country, beginning 1 September, 1993.
5   Reducing the excise tax on unleaded gasoline so that the retail price of a litre of unleaded gasoline would be 0.30 baht (about US$0.012) cheaper than leaded gasoline.[2] The reduction in excise tax revenue would be compensated through the Domestic Gasoline Fund.
6   Disseminating information to the public on the benefits of using unleaded gasoline and to inform automobile users on whether their vehicles can use unleaded gasoline.

Phase II
1   Introduction of a lower excise tax for unleaded gasoline compared to that for leaded gasoline.
2   Regulating the specifications of a fuel system to match international standards and to include a label indicating 'unleaded gasoline only' on the fuel system of each automobile using unleaded gasoline. Requiring that user manuals indicate that unleaded gasoline may be used in the automobile.
3   Regulating automobile exhaust systems and inspecting exhaust emissions.

ENVIRONMENTAL IMPACT

A study by the Thailand Development Research Institute (Tienchai et al, 1990) found that among the various sources of air pollution, automobile emissions comprised by far the largest single source, accounting for 80 per cent of Bangkok's air pollution. The study found that in 1987 some 384 tonnes of lead were emitted, equivalent to a daily emission of about 1,052 kg. From 1988 to 1991, atmospheric lead levels in Bangkok rose and then began to fall (see Table 7.1). This decline can be attributed to the switch from leaded to unleaded gasoline.

A survey conducted by the Ministry of Commerce indicates that in 1992 the consumption of unleaded gasoline amounted to 5.5 million litres compared to 1701 million litres of leaded gasoline. Between January and June 1993, unleaded gasoline consumption rose from 27.4 per cent to 31.8 per cent. By September 1993, consumption of unleaded gasoline rose to 700 million litres and that of leaded gasoline dropped to 1548 million litres. The consumption of unleaded gasoline thus amounted to about 30 per cent of total consumption, while as many as 87 per cent of automobiles were equipped to use unleaded gasoline at that time.

Table 7.1 *Atmospheric Lead Concentrations in Bangkok (micrograms/m³)*

Monitoring Site	1988	1989	1990	1991	1992	1993
Pratunam	1.75	1.97	2.06	1.76	0.66	0.68
Yaovaraj	3.01	2.33	2.21	2.34	0.71	0.61
NSO*	1.38	1.85	4.19	0.94	0.74	0.33
Bamrungmuang	2.29	3.34	5.09	1.92	0.37	–
Sukhumvit	1.06	1.71	–	1.06	–	–
Banglumpoo	0.91	1.15	1.31	1.11	–	0.37
Phaholyothin	0.72	1.18	0.86	0.62	0.94	0.35
Silom	1.95	3.14	2.73	1.90	0.65	–
Si Phraya	0.95	2.81	1.25	1.39	0.54	0.68

Note: * National Statistical Office
Sources: Pollution Control Department, Ministry of Science, Technology and Environment;
Department of Health, Ministry of Public Health

Maneerat (1991) conducted a survey of people's attitudes towards different types of gasoline and found that the reason people used unleaded gasoline was because of its environmental benefits. The study also found that some people continued using leaded gasoline because of the lack of understanding of the effects of different types of gasoline.

Pradit (1992) surveyed 640 people from eight provinces in Thailand and found that 26.3 per cent of people were using unleaded gasoline, 69 per cent leaded and 4.7 per cent use both interchangeably. As many as 66 per cent of those using unleaded gasoline had switched to it because of its environmental benefits. Those who continued to use leaded gasoline thought that unleaded gasoline would reduce the efficiency of the automobile engine. The study also found that unleaded gasoline was used mainly in automobiles with a fuel injection system, while many automobiles with carburettors still continued to use leaded gasoline. Furthermore, the study found that automobile users from Bangkok were more aware of the environmental benefits of unleaded gasoline than those living outside Bangkok.

Nitichai (1994) surveyed 315 people in Chiang Mai City and found that among a set of 16 variables, the price differential had the least effect on consumer decisions regarding the choice of gasoline. Table 7.2 shows the logistic regression results of all 16 variables. Independent variables are dummies which take the value of either one or zero, with 1 indicating a 'yes' answer and 0 a 'no'. Coefficients with positive signs indicate that the presence of these variables will increase the probability of a person using unleaded gasoline.

The results indicated that the price differential had the lowest effect (3.9807) in influencing the probability of a person choosing unleaded gasoline. The study concluded that the major reason people switched to unleaded gasoline was environmental, while those who continued to use leaded gasoline did so due to doubts over unleaded gasoline's effect on engine performance.

Table 7.2 *Factors Influencing the Probability of Using Unleaded Gasoline*

Independent Variables	Coefficient	Significance Level
Constant	−13.8404	0.0181
Plans to use unleaded in the future	23.7071	0.0073
Has a savings account	36.0912	0.0085
Does not believe in negative effects of unleaded	20.7929	0.0078
Self car maintenance	−13.7272	0.0077
Has an interest in using unleaded when informed	−26.3559	0.0073
Male	−21.5240	0.0076
Has knowledge of mechanics	29.9541	0.0080
Executive position at work	−8.3902	0.0071
Hire-purchase payment method	−17.3849	0.0095
Located in municipality	16.6974	0.102
Received less than university education	17.5449	0.0098
No attempt to solve mechanical problem	−14.7624	0.0085
Non working car owner	−24.3871	0.0096
Agrees with 'ban on leaded gasoline'	15.1023	0.0094
Currently not using carburettor system	−15.7962	0.0170
Has an intention to use unleaded if it is cheaper	3.9807	0.0256

Notes: Model chi-square = 367.7180; Goodness of Fit = 36.4850
Source: Nitichai, 1994

ECONOMIC IMPACT

The rationale behind the pricing scheme lies in the negative externality effect of the use of leaded gasoline. As the use of leaded gasoline in automobiles increases the amount of lead in the air, it will eventually lead to adverse effects on public health. To internalize this negative externality, a product tax equal to the marginal damage of using leaded gasoline can be levied on leaded gasoline, thus making it more expensive than the unleaded gasoline. This tax differential would aim to reflect the marginal damage that leaded gasoline has caused to the public. However, since raising the price of leaded gasoline can have a negative effect on the overall economy, the Thai Government decided instead to implement a tax differential by lowering the excise tax on unleaded gasoline by 1 baht (about 4 US cents) per litre. At retail stations the price differential is usually less than 4 US cents per litre. For instance, on 6 January, 1996, the average price of leaded gasoline sold in Bangkok was 35.48 US cents per litre while that of unleaded was 34.28 US cents a litre, indicating that at retail stations the price difference could be as little as 1.2 US cents (0.30 baht) per litre. This smaller price differential at retail stations reduced the potential impact of the gasoline price differential.

For an efficient implementation of the price differential, the government would need to evaluate the marginal damage of a litre of leaded gasoline consumption and set the price differential in accordance with this marginal damage. This method of setting price differential would ensure that those who choose to use leaded gasoline are paying the price of environmental damage. When the price differential is not determined in this fashion, one cannot guarantee that the use of leaded gasoline will decrease to the level considered economically efficient.

A World Bank report (Hammer and Shetty, 1995) examines the case of Malaysia and claims that in order for the leaded gasoline tax to reflect the marginal damage it creates, the tax would need to be as high as 80 per cent of the current price of gasoline. In Indonesia, according to the same report, the tax would need to be as high as 33 per cent. The report adds that at these tax rates there should be no demand for leaded gasoline and all the users will switch to using unleaded. For this reason the report suggests that regulation rather than a gasoline tax should be used to solve the leaded gasoline externality.

The effectiveness of the gasoline price differential is also limited by the inability of some consumers to substitute away from leaded gasoline, since their automobiles have carburettor systems that require leaded gasoline. If unleaded gasoline is used in automobiles that have been designed to use leaded gasoline, it can damage the engine. This limitation prevents some consumers from altering their gasoline consumption behaviour, which in turn further dilutes the effect of the gasoline price differential.

At present, Thai law states that automobiles registered after 1993 are required to use unleaded gasoline only. This change in the technical speci-fication of the automobiles indirectly forces consumers to use unleaded gasoline. In this regard, implementing the price differential may be unnecessary and lowering the price of unleaded gas would then reduce government revenue from gasoline excise tax unnecessarily.

Although the gasoline price differential accords with the Polluter-Pays Principle it tends to produce an undesirable effect on income distribution. The reduction in the price of unleaded gasoline tends to favour high-income automobile owners more than the lower income automobile owners. Nitichai (1994) finds that most of the people who benefit from lower unleaded gasoline price are those whose automobiles were bought after 1991 equipped with fuel injection systems and priced above US$20,000. These people will benefit from the lower price of gasoline as their automobiles require the use of unleaded gasoline. If about 1,548 million litres of unleaded gasoline were purchased in 1993 this would amount to around 464 million baht of implicit subsidy to unleaded gasoline users. Lower income owners of cars bought before 1991 and equipped with the old carburettor system will mostly continue to use leaded gasoline at a higher price.

CONCLUSION

The Thai Government has employed the price differential system in an effort to encourage the public to use unleaded gasoline instead of leaded.

Many automobile users have switched from using leaded to unleaded gasoline but their reason for doing so is not so much the price differential as their awareness of the environmental benefits involved, and the fact that their automobiles are equipped with fuel injection systems which require the use of unleaded gasoline. Other automobile users continued to use leaded gasoline even though it is more expensive.

These observations tend to indicate that the price differential per se in Thailand has done little in terms of altering consumer behaviour towards using unleaded gasoline. Lowering the excise tax on unleaded gasoline has therefore incurred an unnecessary loss of government revenue. Furthermore, lowering the price of unleaded gasoline tends to benefit wealthier automobile owners, while the poor automobile owners, having older cars with carburettor systems, continue to pay a high price for using leaded gasoline.

A slow shift in the regulations on automobile specifications (ie the use of a command and control approach) will prove to be a more efficient method of increasing the use of unleaded gasoline and combating lead pollution, compared to the use of a market-based economic instrument, such as the price differential.

NOTES

1   National Institute of Development Administration, Thailand.
2   Approximately 25 bahts = US$1.

REFERENCES

Tienchai Chongpeerapien, Somthawin Sungsuwan, Phanu Kritiporn, Suree Buranasajja and Resource Management Associates (1990) *Energy and Environment: Choosing the Right Mix*, Thailand Development Research Institute, Bangkok.

Hammer, J S and Shetty, S (1995) East Asia Environment: Principles and Priorities for Action, World Bank Discussion Paper, The World Bank, Washington DC.

Maneerat Ing-orn (1991) Leaded Gasoline Research Report. MSc Thesis, Chulalongkorn University, Bangkok (in Thai).

Nitichai Dawboot (1994) Statistical Analysis of the Unleaded Gasoline Usage at Amphoe Muang Chiang Mai. MSc Thesis, Chiang Mai University, Chiang Mai (in Thai).

Pradit Sririparntong (1992) Behaviour and Attitude of Unleaded Gasoline Users from Non-Bangkok Provinces. *The Research and Development of the Petroleum Authority of Thailand Newsletter*, Bangkok: 19–91.

Suwanna Ruangkanchanasetr and Chamaiphan Santikarn (1994) Lead Poisoning: a Severe Threat to the Nation's Health. *Thailand Development Research Institute Quarterly Review* vol 9 (1), TDRI: Bangkok.

8

GROUNDWATER PRICING IN THAILAND

Adis Israngkura[1]

BACKGROUND

Groundwater has been used in Thailand for many years for both household and industrial consumption. In Bangkok and adjacent areas, records show that groundwater utilization began as early as the 1900s. Groundwater from privately owned wells has been important for household consumption particularly in areas where piped water is not available. Low extraction costs have also made groundwater an attractive source of water supply to the industrial sector. In years of surface water shortage, government agencies such as the Metropolitan Waterworks Authority and the Provincial Waterworks Authority rely on groundwater extraction to meet the deficit.

Uncontrolled utilization of groundwater has led to environmental problems such as land subsidence and groundwater depletion. Open access to groundwater utilization continued until around 1977 when the Department of Mineral Resources (DMR) began to control the use of groundwater. Economic instruments to control the use of groundwater did not begin until 1984 when DMR imposed a user charge of 1 baht (US$0.04) per cubic metre of water. As land subsidence due to over-extraction of groundwater continued, this user charge was raised to 3.5 baht (US$0.14) per cubic metre in 1995. In areas where piped water is unavailable, a 25 per cent discount was applied.

The main problem associated with groundwater extraction is land subsidence, which is particularly serious in the Bangkok Metropolitan Area, which comprises Bangkok metropolis and the neighbouring provinces of Nakhon Pathom, Nonthaburi, Pathum Thani, Samut Prakarn and Samut Sakhon. Land areas have been grouped into different classes based on the severity of land subsidence. In critical area Class I (highly critical) the rate of land subsidence is as high as 10 centimetres or more a year. The Class I area includes 11 districts in Bangkok, two districts in Pathum Thani province and three districts in Samut Prakarn province. The critical area Class II (moderately critical) where subsidence is between 5–10 centimetres per year includes five districts in Bangkok, three districts in Samut Prakarn province, two districts in Nonthaburi province, three districts in Samut Sakhon province, six districts in Pathum Thani province and two districts in Nakhon Pathom province. Lastly, critical area Class III covers areas with land subsidence of less than five centimetres per year; these include the remaining districts in Bangkok, Nonthaburi, Pathum Thani and Samut Prakarn.

In addition to land subsidence, groundwater depletion also raises the possibility of saltwater intrusion. Some of the critical provinces discussed above are located near the Gulf of Thailand, where seawater contamination of groundwater reservoirs is likely. Although this problem has not occurred widely, further extraction in some of these areas could increase the probability of saltwater intrusion.

In terms of resource utilization, excessive use of groundwater beyond its natural recharge rate would eventually deplete the stock and further aggravate the existing environmental problems. Such excessive use would thus impose a cost on future generations, as resources are taken from future stores for present consumption. This cost is known as the user cost of groundwater extraction. If the benefit of groundwater utilization today is less than this user cost (ie the forgone future benefit), groundwater utilization would be considered sub-optimal and would result in a loss in social welfare.

Groundwater extraction in the Bangkok Metropolitan Area and the adjacent provinces totals 1.468 million m^3/day, not including extraction by government agencies and state enterprises. Of this, 81.4 per cent is used by the industrial sector and only 17.9 per cent by households. Industrial use of groundwater is concentrated mainly in Samut Prakarn and Pathum Thani provinces, where many industrial establishments are located.

Groundwater extraction is common throughout Thailand, but particularly in areas where piped water is unavailable. Rural households and those not served by the Provincial Waterworks Authority (which provides piped water) usually resort to installing water pumps to extract groundwater for consumption; many poor villagers who cannot afford water pumps rely on well water. These small users usually operate water pumps without permits from the DMR, and the extraction is therefore not controlled by the authorities. Many small users find that after several years of extraction they have to increase the depth of extraction as the level of groundwater drops. This indicates that although only some areas are suffering land subsidence, groundwater depletion is a common problem throughout Thailand.

Based on the groundwater concession records issued by the DMR, groundwater extraction in Thailand (including both private and public users) amounts to 1.804 million m³/day covering a total of 7,595 wells. Public agencies extract about 400,000 m³/day. Of this, the Metropolitan Waterworks Authority extracts around 250,000 m³/day, the Provincial Waterworks Authority 68,229 m³/day and other agencies around 80,000 m³/day.

It has been estimated that water utilization in Thailand will almost double in the next 20 years, from 5.3 million m³/day in 1995 to 9.7 million m³/day in 2017. Much of this increase is expected to come from the rising demand for water from the industrial sector. However, due to the expected expansion of the piped water system, it has been estimated that groundwater utilization will stabilise at around 1.4 to 1.6 million m³/day over the next 20 years. From this estimation industries will account for about 70 per cent of the total groundwater use.

GROUNDWATER PRICING

To combat the excessive use of groundwater in Thailand the DMR adopted an economic instrument known as groundwater pricing. In 1984, the user charge for groundwater was set at the rate of 1 baht (US$0.04) per cubic metre. In addition, from 1987 the Metropolitan Waterworks Authority banned groundwater extraction in Class I and II areas. Despite these measures, land subsidence continued and in February 1995 the user charge was increased from 1 baht to 3.5 baht (US$0.14) per cubic metre. For areas where piped water is unavailable, a 25 per cent discount was provided to commercial businesses, industries and farming, while households in these areas were totally exempted from the user charge. This pricing scheme was expected to increase the cost of extraction and hence decrease extraction rates.

Prior to 1984, groundwater in Thailand was accessible free of charge to practically all users. Although the Groundwater Act of 1977 requires that all users obtain a permit from the DMR for installing water pumps and extracting groundwater, very little effort was made to regulate the volume of groundwater extracted. Small users such as individual households often managed to install water pumps without having to obtain permits from the DMR and thereby extracted groundwater free of charge.

This open access has led to inefficiencies in groundwater utilization. When groundwater is available free of charge, users will not realize the forgone benefits or the opportunity costs of using this resource. Groundwater utilization involves three types of opportunity cost. First, if groundwater is exploited by one user, this will reduce the amount available to others. In this case, using groundwater will involve some forgone benefits and will have an 'opportunity cost'. Second, if groundwater utilization today exceeds its natural recharge rate, it will reduce the water stock, reduce groundwater availability for future consumption and so reduce future benefits from groundwater. This inter-temporal opportunity cost of using groundwater is known as the 'user cost'. And, third, if groundwater extraction causes land subsidence and hence increases the frequency of flooding and other environmental damages (such as the possibility of seawater intrusion) groundwater extraction will involve some 'environmental costs'. Combined, these three types of opportunity cost constitute the total opportunity cost of groundwater extraction, which is also known as the social cost of groundwater extraction.

For groundwater to be utilized efficiently, users must take into account this social cost. To attain economic efficiency, groundwater extraction should reach the level at which the benefits of the last unit of water extracted equals the social cost of extraction. In other words, economic efficiency will be realized when groundwater is extracted until the marginal benefit equals the marginal social costs of doing so. One way to attain this economic efficiency in groundwater extraction is to impose a price or user charge on groundwater at the rate equal to the marginal social costs. When the user charge is set in this manner, water users will have an incentive to limit their use. Groundwater extraction for an individual will continue until the marginal benefit from using groundwater equals this user charge (marginal social costs).

By making groundwater more expensive, the user charge for groundwater is expected to provide some incentives to divert water users to other water sources and hence reduce groundwater extraction. Imposing a user charge on groundwater will not only provide incentives to water users to search for other water sources, but it will also encourage them to become more conservative about groundwater use. As every unit of groundwater extracted will incur cost, water users will have an incentive to fix leakages, reduce wasteful use or search for a more efficient method of water utilization. Any water saved through these efforts will translate into a reduction in groundwater bills.

IMPLEMENTATION ISSUES

Thailand has little experience in using economic instruments to regulate natural resource use. Although resource pricing has been applied in forest and mineral concessions in the form of royalties, the objective of such pricing has been mainly to generate government revenue rather than to increase economic efficiency in resource utilization. In the case of the user charge on groundwater extraction, the objective of pricing groundwater is not merely to raise revenue but rather to reduce the rate of groundwater extraction.

Groundwater extraction has been regulated since 1977 when the Groundwater Act 1977 came into effect. At that time groundwater was allocated under a command and control regime, that is, by issuing groundwater permits. From 1977 onwards, groundwater users needed to obtain permits from the DMR before extracting groundwater. These permits, which are transferable, include drilling permits, extraction permits, discharge permits, and so on.

It was also stated in the Groundwater Act of 1977 that groundwater permits can be revoked if extraction results in any damage to the groundwater reservoir, stock of water, environment or public health, or leads to land subsidence.

Under this regime the DMR did not require that water meters be installed at water pumps and hence the DMR was unable to control the volume of groundwater extracted. Furthermore, numerous small households extract groundwater for their own private consumption and these activities are practically beyond the control of the DMR. For these reasons, groundwater extraction continues to escalate, causing severe land subsidence in many areas.

After introducing the groundwater user charge in 1984, the government took a further step in 1987 by prohibiting the Metropolitan Waterworks Authority from using groundwater. The government also announced a total ban on all uses of groundwater in critical area Class I (land subsidence of 10 cm or more per year) from January 1997. From 1998, the ban has been extended to the critical area Class II (land subsidence between 5–10 cm per year). Furthermore, the DMR will stop issuing groundwater permits and revoke permits in areas where piped water has become available. This total ban on groundwater is expected to help eliminate land subsidence problems in critical areas.

Groundwater extraction is regulated by four main bodies of legislation and regulations: the Groundwater Act 1977, the Groundwater Act (Amendments) 1992, various Ministerial Regulations and Announcements of the Ministry of Industry, and Announcements of the DMR.

The revenue collected from groundwater utilization is administered by the DMR. Currently, record keeping of the volume of groundwater extraction is not based on water meter readings. Instead, each user is required to report to the DMR every three months on the amount of water extracted. The DMR calculates the water bill using this reported volume and comparing it with the allowable amount stated on each permit and the type of business/activity in operation. Due to the shortage of staff, the DMR sends out officials to check the quantity of water actually used only if the reported quantity is markedly different from the volume which would be expected from the type of business involved.

IMPACT

According to the Groundwater Act of 1977, the user charge for groundwater cannot exceed the user charge for piped water, currently set at around 7 baht (US$0.28) per cubic metre. This stipulation prevents the user charge for groundwater from reflecting the true marginal social cost of extraction and hence efficient resource utilization.

Assessing whether the user charge on groundwater will have a desirable distribution effect will require further empirical testing. Although a business owner without access to piped water can obtain a 25 per cent discount, it is not clear how this benefit will be distributed among consumers and producers.

The Groundwater Act of 1977 requires that users must obtain permits from the DMR before extracting groundwater. The Act also allows these permits to be transferred. Given this situation, the DMR might consider developing this permit system into one of tradable permits for groundwater extraction. However, an effective tradable groundwater permit system would require two additional measures. Firstly, the DMR would need to set an upper limit on the number of permits issued in each area, making these permits tradable among users. Secondly, the number of permits issued should differ from one area to another depending on the degree of severity of land subsidence or groundwater depletion. Once this number is determined, the 'price' of the permits could be established by the market itself and reflect the opportunity cost of groundwater use.

Effective implementation of the groundwater user charge currently faces two limitations: a lack of enforcement and the inability to control the volumes extracted. When the volume of extraction is not closely monitored, the user charge imposed will not function properly. By understating the true volume of groundwater used, users may simply lower their water bills and hence reduce the effectiveness of the user charge.

CONCLUSION

The Thai Government has attempted to control excessive use of groundwater through resource pricing, namely by imposing a user charge on groundwater extraction. This method is used together with the command and control method where a complete ban on groundwater extraction is effective in critical areas. In order for the system to function more efficiently there is a need to revise some of the laws and regulations that have prevented the price of groundwater from reflecting its marginal social costs. The latter includes not only the opportunity costs but also the user cost and environmental costs. Making groundwater permits tradable is another measure that would greatly benefit Thai society and ought to be explored. As groundwater pricing is still very new to Thailand, one can only wait and see its impact, particularly in critical areas where land subsidence is pervasive.

NOTE

1   National Institute of Development Administration, Thailand.

9

THAI INDUSTRIAL ESTATE WATER TREATMENT FEE

Adis Israngkura[1]

BACKGROUND

Industrial estates have been developed in Thailand, as elsewhere, to confine factories or industries to a given area. This method of clustering factories together so that they operate within a given boundary benefits society in many ways. Firstly, an industrial estate can act as a zoning device for the city so that factories will not create negative externalities on other activities and land uses, such as commercial centres or residential areas. Secondly, as output from one factory may be used as input for another, locating these factories in close proximity to one another will help reduce transportation costs and other kinds of transaction costs. And thirdly, as most of these factories tend to produce by-products such as wastewater, developing a central wastewater treatment system to serve all the factories in the same locality tends to produce an economy of scale and thus helps reduce the costs of pollution control.

Industrial estates are becoming increasingly important to Thailand's industrialization. By 1995 there were 23 estates, including both public and private ones (see Table 9.1). Their wastewater treatment capacities range from 2,500 m³/day to 23,700 m³/day.

Factories operating within an industrial estate are provided with a central wastewater treatment facility. Before the wastewater is sent to the central treatment system it has to be pre-treated within the factory so that the pollution load measured in terms of BOD (biochemical oxygen demand) does not exceed 500 mg/l and suspended solids do not exceed 200 mg/l.

Factories are charged a fee for sending their pre-treated wastewater to the central treatment system. This fee is equivalent to a user charge on wastewater released by the factories. Public industrial estates set the user charge to cover the cost of operating the wastewater treatment facilities. This method of pricing is also known as average cost pricing.

Table 9.1 *Wastewater Treatment Systems and Capacities of Thai Industrial Estates*

Industrial Estate	Treatment System	Capacity (m³/day)
1 Bang Chan	na	8,000–10,000
2 Bang Plee	Activated sludge	21,000
3 Bang Pu	Aerated lagoon	23,000
4 Gemopolis	Activated sludge	2500
5 Lad Krabang	Activated sludge	15,700
6 Samut Sakhon	Activated sludge	21,000
7 Bang Pa-In	Activated sludge	12,000
8 Bang Pra Kong	Aerated lagoon	6,000
9 Chonburi (Bor Win)	Activated sludge	8400
10 Gateway City	Activated sludge	na
11 Hi Tech	Activated sludge	16,800
12 Nong Kare	Activated sludge	12,000
13 Saharat Nakhon	Activated sludge	8600
14 Sara Buri	Activated sludge	8800
15 Welco	Aerated lagoon	16,000
16 Eastern	Activated sludge	12,000
17 Leam Chabang	Activated sludge	23,700
18 Mab Ta Phut	Activated sludge	4000
19 Northern	Aerated lagoon	5600
20 Pha Deang	na	12,000
21 Pichit	na	5100
22 Songkla (Chalung)	Activated sludge with extended aeration	3000
23 Udon Thani	na	na

na = data not available
Source: Industrial Estate Authority of Thailand

WASTEWATER TREATMENT FEE

The wastewater treatment fee is derived using the following principles. Firstly, the fee is collected from the polluters, which is in accord with the Polluter Pays Principle. The fee is calculated on the basis of the quantity and pollution load of the wastewater. Secondly, factories are required to pre-treat the wastewater before sending it to the central treatment facility. The pollution load in the wastewater before it enters the central treatment facility must not exceed 500 mg/l of BOD and 200 mg/l of suspended solids. Thirdly, as the wastewater treatment systems vary from one industrial estate to another, a general formula (given below) outlines the common variables to be used in calculating the fee. The specific values of the parameters used in the formula will vary according to the cost structure of each industrial estate. Fourthly, as many wastewater treatment systems are designed to treat organic waste, the formula is derived based mainly on the pollution load measured in terms of BOD. Suspended solids are not directly part of the general formula but enter the equation as an excess fee or penalty cost.

Each industrial estate follows the general formula in calculating the wastewater fee although the specific parameters used may be different depending on the actual cost of treatment which varies from one industrial estate to another.

The general formula is given as:

$$TC = Cg + Cf + Cv + Cp$$

where:
TC = total cost (baht per month).
Cg = general cost; including administrative costs or wastewater testing fee.
Cg = K0
Cf = fixed cost, including depreciation costs of the capital equipment and O&M independent of BOD removal.
Cf = ViK1
Cv = variable cost; includes O&M costs dependent on BOD removal.
Cv = (ViSi/1000)K2
Cp = penalty cost; which is a fee imposed when wastewater quality is below the standard specified by the estate.
Cp = ViK3 when suspended solid exceeds 200 mg/l or
Cp = 3[Cg+Cf + Cv] when other pollutants exceed standards set by the estate.

Thus:

$$TC = \quad K0 + ViK1 + (ViSi/1,000)K2 + ViK3$$

where:
Vi = quantity of wastewater from the factory (m³/month)
Si = water quality measured in terms of BOD_5 (mg/l)
K0 = US$4 per month (baht100/month).

$$K1 = \frac{1/10 \text{ (pipe system cost)} + 1/15 \text{ (treatment system cost)} + 10/100 \text{ (total construction costs)}}{(12)(30)(\text{Wastewater Treatment Capacity})}$$

$$K2 = \frac{15/100 \text{ (Total Construction Costs)}}{(12)(30)(\text{Wastewater Treatment Capacity}) \times B/1000}$$

where:
B = BOD_5 standard for the effluent set by the industrial estate

K3 depends on the amount by which the quantity of suspended solid (SS) exceeds the standard set at 200mg/l. If SS exceeds this standard then the following values of K3 are used:

K3 = US$0.08/m^3/month (2 baht/m^3/month) if 200 < SS < 400.
K3 = US$0.16/m^3/month (4 baht/m^3/month) if 400 < SS < 600.
K3 = US$0.32/m^3/month (8 baht/m^3/month) if 600 < SS < 1,000.

Note: In K1 above, the fractions 1/10 and 1/15 represent the annual depreciation costs of the piped system costs and treatment plant costs, which are assumed to last for 10 and 15 years, respectively. The fraction 10/100 represents the share of the total construction costs (pipe system and treatment plant), calculated at 10 per cent. The numbers 12 and 30 in the denominator represent months in a year and days in a month, and are used for calculating annual wastewater capacity.

Similarly, in K2 above, the fraction 15/100 represents the share of the total construction costs (pipe system and treatment plant), calculated at 15 per cent or 15/100. As in K1, the numbers 12 and 30 are employed to calculate annual wastewater capacity.

Examples:
1   At the Laem Chabang Industrial Estate the total cost of wastewater treatment per month is calculated as:

$$TC = 100 + 6.10Vi + 10.47ViSi/1{,}000 + ViK3$$

2   At the Mab Ta Phut Industrial Estate the total cost of wastewater treatment per month is calculated as:

$$TC = 100 + 6.12Vi + 10.75ViSi/1{,}000 + ViK3$$

When the industrial estates are equipped with central treatment facilities and charge a user fee for treating wastewater it is expected that the cost of wastewater treatment will be reduced. The total treatment cost is expected to be lowered because this system recognizes that individual factories are more cost efficient in treating wastewater with a high BOD load and the central treatment facility with an economy of scale is more cost efficient when treating wastewater with a lower BOD load.

At a lower pollution load, namely when BOD$_5$ is below 500 mg/l, the cost of further treatment by an individual factory will increase tremendously and treating at a central treatment system will become more cost saving. Thus, having the factories pre-treat their own wastewater before sending it to the central treatment system helps minimize the cost of pollution control.

DESIGN AND IMPLEMENTATION ISSUES

Wastewater released from factories which are not located in industrial estates are subject to end-of-pipe wastewater standards or effluent standards. This command and control method has proven to be expensive for the factories as it does not exploit the cost advantage of a central treatment facility. The availability of a central wastewater treatment facility may be one of the reasons why many factories choose to locate inside industrial estates, as they can enjoy this cost advantage.

Application of a user charge is also recognized in the Industrial Estate Act of 1979 which allows industrial estates to impose a 'fair' user charge for any services provided.

The fee is based on the average cost concept and not on the marginal cost concept. This average cost pricing technique will allow the industrial estate to cover the cost of wastewater treatment (break-even) but it will usually mean that allocation efficiency will not be realized. In order to efficiently allocate resources for pollution control, wastewater should be treated until the marginal cost equals the marginal benefit. For private industrial estates, using the concept of average cost pricing may not create much economic inefficiency as their investment is still oriented toward profit maximization. With public industrial estates, however, using the average cost concept to determine the user charge could lead to excess capacity as the estate can pass on all costs to the factory owners.

In determining the user charge on wastewater treatment there has been no attempt to measure the marginal value (or benefit) of wastewater treatment. When this information is not known or is not used to calculate the user charge, treatment is probably being carried out at a sub-optimal level. The pre-treatment method will, however, ensure that the cost of treating wastewater will be minimized as the factories are able to treat wastewater more cheaply when the pollution load is high and the central treatment system can treat wastewater more cheaply when the pollution load is low.

CONCLUSION

Industrial estates play an important role in Thailand's industrialization by providing factories with necessary infrastructure and by internalizing negative externalities. Imposing a user charge on wastewater treatment and requiring that wastewater be pre-treated helps reduce the cost of pollution control. However, the marginal benefits of treating water should be taken into account and this information should be used in determining wastewater standards and/or in calculating the user charge of wastewater treatment.

NOTE

1   National Institue of Development Administration, Thailand.

REFERENCES

Adis Israngkura (1995) Economic Instruments in Wastewater Management, in
    Kaosa-ard, M and Israngkura, A (eds), *Frontier Knowledge on Water Issues
    in Thailand*, revised second edition, Thailand Development Research
    Institute, Bangkok (in Thai).

10

ENTRANCE FEES FOR USERS OF KHAO YAI NATIONAL PARK, THAILAND

Mingsarn Kaosa-ard and Eric Azumi[1]

BACKGROUND

The acreage of protected land has grown rapidly in Thailand. Between 1987 and 1992, the amount of land protected by the National Park Act (1961) increased by more than 40 per cent. However, due to budget constraints, funding for the maintenance and continued protection of these areas has not risen at a commensurate rate. At Khao Yai National Park in particular, the costs of park maintenance have risen 15 to 20 per cent annually in recent years, while cost recovery over the same period has dropped from 51 to 40 per cent (Table 10.1).

This case study examines the possible consequences of raising the entrance fees to Khao Yai National Park as a method of capturing more of the public's willingness to pay (WTP) for both the use and continued existence of the park. If the increase in fees can raise enough revenue to cover the costs of park maintenance and protection, it thus ensures that those who value and use the park most are paying for its upkeep, a more efficient and sustainable method of financing the costs of environmental conservation.

Khao Yai is Thailand's oldest and most popular national park and covers an area of 2168 km^2. It receives nearly a million visitors a year, a little less than 10 per cent of the total number of visitors to national parks throughout the country. Much of its popularity can be attributed to its proximity to Bangkok; Khao Yai is a three-hour drive from the city.

Environmentally, Khao Yai is quite significant. It is abundant with precious plants and animals and has been classified as an ASEAN Heritage Park. Over 2000 species of plants and 294 species of birds can be found inside the park boundaries. Researchers from around the world

Table 10.1 *Costs and Revenue for Khao Yai National Park (US$; US$1 = 25 baht)*

Items	1991	1992	1993	1994	1995
Revenue					
Operating revenue:	174,838.28	168,478.40	150,692.92	192,380.40	na
entrance fees	143,543.88	149,217.80	138,065.92	166,637.00	na
fines	16,487.20	13,990.00	11,412.00	8,270.00	na
accommodation	14,807.20	5,270.60	1,215.00	17,473.40	na
Allocated Budget:	262,256.00	347,440.00	458,868.40	484,720.00	793,280.00
Total park revenue	437,094.28	515,918.40	609,561.32	677,100.40	na
Costs					
Cost of protection	232,108.68	283,234.52	322,557.60	310,914.20	352,245.76
Cost of tourism service	115,611.96	141,743.84	169,598.80	169,554.64	184,588.52
Road maintenance	–	–	–	–	321,309.12
Total cost	347,720.64	424,978.36	492,156.40	480,468.84	861,743.40
Increase in cost (%)	–	22.22	15.81	–2.37	11.73*
Cost recovery (%)	50.28%	39.64%	43.34%	40.04%	na
Number of visitors	1,009,687	944,940	729,818	817,261	na
Cost per tourist	0.34	0.45	0.67	0.59	na
Average fee per tourist	0.17	0.18	0.21	0.24	na

Notes: na = information not available; * = does not include costs of road maintenance
Source: Khao Yai National Park

travel to Khao Yai to study elephants, hornbills, and gibbons. It is also the head watershed of three major river systems, and provides watershed services to four provinces in the region, with a combined estimated discharge of 1,889 million m^3 per year.

Despite the myriad of services that Khao Yai provides to both Thailand and the world as a whole, these benefits are rarely measured, largely because of the difficulty of putting a monetary value on the non-market goods and services that a national park provides. In a recent study, the Thailand Development Research Institute (TDRI), using two methods – contingent valuation and travel cost – surveyed 1057 non-users and 948 users of the park in order to determine the general public's willingness to pay for the various uses, both direct and indirect, of Khao Yai (TDRI, 1995).[2] This type of valuation is one method of estimating the full value of a natural resource such as the park, and provides policy makers with a measurable indication of the benefits that the park provides to the nation as a whole. Results of the surveys are given in Table 10.2.

Table 10.2 *Average Annual WTP for Khao Yai National Park*

	Non-users	Users
Average WTP	183 baht	730 baht
Method used	Contingent Valuation	Travel Cost
Number of respondents	1056	763

Note: 25 baht = US$1
Source: TDRI, 1995

The TDRI report aggregates the combined benefits for both park users and non-users, based on various categories of respondents (eg, income level, distance of home from the park, etc.), for the country as a whole. The report concludes that 'the total recreational, environmental and biodiversity benefits of Khao Yai National Park, for both park users and urban non-users, total US$108.2 million per year.' The question remains, though, how to translate these benefits into revenue that will help finance continued protection of the park.

POSSIBLE INCREASED ENTRANCE FEE

The method that TDRI recommends, and which this case study examines, is raising the cost of entering the park. Currently, the average entrance fee to the park is US$0.20 per person. This rate has not changed in over ten years. TDRI estimated that the cost of providing services to tourists (including wilderness protection) was about US$0.60 per person in 1994 (Table 10.1). Thus, those who benefit most directly from continued park protection are paying only one third of its cost.

The budget allocated by the government is barely enough to maintain basic services. Often maintenance of buildings is forgone, which results in a general deterioration of park facilities. In addition, there is virtually no funding for research into the biodiversity and forest ecology of Khao Yai. In 1988, only US$500 was allocated for research in Khao Yai. The park has no plant or wildlife specialists. The value of Khao Yai as a research and educational centre is not fully realized, due largely to this lack of personnel and facilities.

Not only does the entrance fee fail to recover fully the costs of managing the park, but it also fails to reflect visitors' willingness to pay. In TDRI's estimation, Thai visitors were willing to pay an average of US$0.88, well above the current entrance fee, as well as the cost of providing services. Of the 946 respondents, 259 (27 per cent) refused to pay any increase. In addition, TDRI surveyed visitors' willingness to pay for improved services (ie, improved roads, facilities, animal viewing sites). The average amount visitors were willing to pay as an entrance fee with improved services was US$1.76, with only 58 of the 946 respondents unwilling to pay for an increase.

Foreign tourists were surveyed as well. Their willingness to pay to enter Khao Yai was significantly higher than their Thai counterparts. The average WTP was US$5 for current levels of service and US$5.72 for improved services.[3]

TDRI also surveyed the use of the park as a thoroughfare. Travelling through the park is a 50 km shortcut for traffic from Prachinburi in the Eastern region to Saraburi, the gateway to the North-eastern region of Thailand. There has been a continuing controversy over the use of the park for through-traffic. Opponents argue that the road is hazardous for much of the wildlife, especially at night. There is anecdotal evidence of deer and other animals being killed by vehicles, but no official records are kept. Proponents of keeping the road open note the economic benefits of the shorter route and lower transportation costs, and point out the disadvantages for those tourists who wish to continue to points on the far side of the park.

A TDRI survey estimated that about 27 per cent of the vehicles that enter Khao Yai National Park use it solely for through-traffic purposes, and most of these are passenger cars and pick-up trucks. The six-wheeled trucks that use the Park as a throughway tend to use it on their return trips when their trucks are empty, as the roads are hilly and winding.

An estimated 3,000 trucks pass through the park every year rather than go around it and these vehicles are responsible for much of the road damage, noise and air pollution in the area. The income from this through-traffic, however, accounts for only three per cent of total entrance fee revenues (Table 10.3).

Table 10.3 *Revenue from Vehicle Entrance Fees*

Type of Vehicle	Entrance Fee Baht	(US$)	Per Cent of Revenue 1991	1992	1993
Bicycles	5	(0.2)	0.02	0.05	0.02
Motorcycles	10	(0.4)	14.68	7.01	3.67
Passenger cars	25	(1.0)	51.59	54.22	43.25
Vans	60	(2.4)	16.41	22.67	16.39
Small coaches	150	(6.0)	7.88	7.27	13.27
Big coaches	200	(8.0)	6.31	6.32	18.92
Trucks	30	(1.2)	3.12	2.43	4.32

Note: Ten-wheeled trucks are not allowed to enter the park
Source: Khao Yai National Park

TDRI estimated the costs of a six-wheeled truck travelling through Khao Yai as opposed to going around (Table 10.4). The savings differential for an empty truck is about US$0.88. For a full truck it is in fact more expensive (247 baht or US$9.88) to travel through the park when one takes into account the wear and tear on the vehicle. For other types of vehicle, the costs of travelling through the park and making a detour are not significantly different.

Thus it seems that through-traffic does not provide major economic benefits to the park nor to the transportation industry. In this regard, TDRI recommends closing the park to through-traffic, or at least to six-wheeled trucks.

Thailand's national parks fall under the jurisdiction of the Royal Forest Department (RFD) of the Ministry of Agriculture and Cooperatives. The Superintendent of the National Park, an RFD employee, has a significant amount of discretion in the daily operations of the park, although the setting of entrance fees and the like requires

Table 10.4 *Cost Comparison between Traffic Through Khao Yai and the Detour*

Item	Through Khao Yai		Detour	
	With goods	Without goods	With goods	Without goods
Wages	8.00	8.00	8.00	8.00
Depreciation	19.44[a]	4.86[b]	6.48[c]	3.24[d]
Petrol[e]	8.61	6.97	11.31	9.04
Entrance fee	1.00	1.00	0	0
Time cost	0	0	1.40	1.40
Total	37.05	20.83	27.19	21.68

Notes: Unit: US$; US$1 = 25Baht
a Assumed working life = 2 years
b Assumed working life = 8 years
c Assumed working life = 6 years
d Assumed working life = 12 years
e Petrol consumption = 4.4 km/litre through Khao Yai, 8.0 km/litre via detour
Source: interviews, 1994

approval at the national level from RFD headquarters. The National Park Act 1961 empowers the Director-General of RFD to collect fees from park visitors. This fee has to be approved by the Minister of Agriculture and Cooperatives. Not all national parks charge entrance fees, especially those closer to urban areas. For the rest, however, fees are set at a uniform rate throughout the country, regardless of differences in the number of visitors, costs of park maintenance and operation, and the attraction of each park. The RFD will have to change its policy of uniform pricing before individual parks are allowed to set entrance fees that more closely reflect the park's circumstances.

There may be other means of capturing more of the benefits of protected areas such as Khao Yai. The educational and scientific benefits are one such opportunity. International researchers who wish to study the extensive flora and fauna could be charged a percentage of their budgets for conducting their research. Dobias (1986, cited in TDRI, 1995) compiled data on the flow of funds related to research in Khao Yai (Table 10.5). While all of the money was not spent in the park, it gives an indication of the amount spent over 12 years to study flora and fauna in the park.

POSSIBLE IMPACTS

If entrance fees to the park were raised from their current US$0.20 to US$0.80 per person, TDRI predicts an estimated 27 per cent drop in the number of visitors.[4] If the remaining 73 per cent of the 817,261 visitors to Khao Yai in 1994 paid US$0.80 per head, Khao Yai would have raised US$477,280 or 99 per cent of its costs for 1994. Moreover, if visitors

Table 10.5 *Scientific Research Funding for Topics Related to Khao Yai (1976–1988)*

Research funding	Baht (1992 prices)	US$
Gibbons	1,713,538	68,542
Hornbills	3,282,283	13,129
Elephants	33,045	1,322
Others	7,541,127	301,645
Total	12,569,993	502,800

Source: TDRI, 1995

perceived that park services were improved, they would be willing to pay an even higher entrance fee.

For foreign visitors, TDRI suggests raising the entrance fee to US$2. Though the mean WTP for foreign visitors was US$5, the mode was only US$2. This differentiated fee is common in some other countries with significant eco-tourism, such as Kenya and Costa Rica.

Raising entrance fees to US$0.80 per person may cause some controversy, especially among those who consider national parks to be public goods. There was a small minority (17 out of 939) of survey respondents who submitted protest bids, stating that they felt the government should subsidize all costs of the national parks.

In addition, the predicted 27 per cent decrease in visitors may have some impacts on tourism-related activities in the area surrounding the park. There are a number of resorts that have sprung up in recent years catering to tourists visiting the park.[5] These resorts, however, are relatively expensive, and an increase of US$0.6 in the cost of entering the park will mostly affect users of the camping facilities, which are free.

There is also revenue that accrues to people from some of the villages surrounding the park who act as porters for treks in the park. TDRI estimated that this income averaged about US$40 per year, a relatively small amount compared to what villagers could earn from extracting resources from within park boundaries. The gazetting of the park area has cut off income to these villages, an issue that needs to be addressed by the RFD.

The higher entrance fee may put increased pressure on other parks near Bangkok, as they may be close substitutes for Khao Yai. In fact, the RFD may want to consider raising entrance fees at all of its parks for the same reasons that are justified at Khao Yai.

CONCLUSION

Khao Yai National Park is currently under-priced. Visitors are willing to pay significantly more than they are currently paying to enter the park, while each year park officials face increasing costs of maintenance. This has meant a lack of maintenance of park facilities, and minimal funds for improvements in park services.

By increasing the price of the entrance fee, Khao Yai could become largely self-financing, a much more efficient and sustainable method compared to relying on annual government budget allocations. Increases in funding would allow for investment in park services, and facilities such as nature centres and libraries could be created and maintained, thus increasing the value of the park to its visitors. The long-term health of the park and its utility to the country would be well served by the proper pricing of access to the park.

NOTES

1   Thailand Development Research Institute (both authors).

2   For a complete description of the methods used and analysis of the results, see TDRI, 1995.

3   In addition to the WTP survey for the entrance fee, a separate survey also investigated the WTP of visitors and non-visitors to conserve the park. This latter study showed that on average Thai visitors were willing to pay 730 baht per year as opposed to 551 baht offered by foreign visitors. For non-visitors, average WTP was 183 baht (Thai) and 121 baht (foreign).

4   This figure is based upon the fact that 27 percent of survey respondents refused to pay a higher entrance fee.

5   Many of the resorts have opened since 1991, when the golf course and the Tourism Authority of Thailand accommodation facilities inside the park were closed. This closure received a significant amount of publicity, which many interpreted as a closure of the park as a whole. This may account for the drop in the number of visitors after 1991.

REFERENCE

Thailand Development Research Institute (TDRI) (1995) *Green Finance: A Case Study of Khao Yai.* Natural Resources and Environment Program, TDRI, Bangkok.

PART III

EASTERN AND CENTRAL EUROPE CASE STUDIES

11

SULPHUR DIOXIDE EMISSION CHARGE IN POLAND

Zsuzsa Lehoczki[1] and Jerzy Sleszynski[2]

BACKGROUND

In the early 1980s there was a general revival of the Polish national environmental policy, largely due to the impact of the Solidarity movement. Air pollution problems had a high profile because of their associated health risks and particular attention was paid to sulphur dioxide (SO_2) emission because of the very obvious environmental damage caused by acid rain, especially in the forests in the southwestern part of Poland.

At the beginning of 1980, the Polish parliament replaced the rather inadequate Nature Protection Act with the first general Legal Act on Protection and Management of the Natural Environment, thus introducing an extended set of environmental charges that could be levied on polluters. These new charges were expected to play an important role in every area of environmental protection. The general act with its numerous amendments also established a complex regulatory environment for SO_2 emission control and an SO_2 charge appeared as a new policy instrument.

However, by the second half of the 1980s, it became clear that the new legislation was not bringing about sufficient changes. Poland's sulphur dioxide emission levels declined by only 10 per cent over the period 1985–1989. In 1992 the SO_2 emission per unit of GDP was still five times higher than that of Germany and about six times higher than the average for European OECD countries (OECD, 1995).

Studies had shown that poor air quality was a serious threat to human health in several cities during the early 1990s. Sulphur dioxide concentrations in cities like Chorzow or Krakow far exceeded ambient air quality standards and were several times higher than in German cities with comparable industrial activities. The national air quality monitoring data indicate that SO_2 concentration violated air quality standards in about 15 of Poland's 29 regions between 1990–1992 (OECD, 1995).[3]

The levels of both the export and import of sulphur dioxide pollution remained high throughout the eighties. By 1992, however, SO_2 export was almost double the amount imported. Nordic countries, the major recipients of the unwanted export, exerted pressure on Poland and offered assistance in reducing the emission levels.

The major cause of Poland's heavy SO_2 pollution has been the heavy reliance on coal-fired power plants for electricity production: in 1992 public power production accounted for almost half of the country's sulphur dioxide emissions. Public and private heating contributed 26.6 per cent, with a further 24 per cent coming from industrial power generation and industrial processes.

The 1991 National Environmental Policy (NEP) assigned a high priority to the need to reduce air pollution. One of the medium term objectives to be achieved by 2000 is a 30 per cent reduction of SO_2 emission from its 1980 level. This will require a reduction in the annual emission from 4.2 million t/year in 1980 down to 2.9 million t/year by the year 2000. A significantly increased SO_2 emission charge has been envisioned as an important policy tool to help reach this target.

Emission Charge

The sulphur dioxide charge is designed as a classic emission charge. It is a price to be paid by polluters for each unit of SO_2 they release into the air. The aim of the charge is to create an economic incentive against excessive air pollution from fuel burning.

The SO_2 charge is closely connected to the system of permits for enterprises that are point sources of air pollution. These permits specify the allowable emission of every regulated pollutant from every stack or other type of conveyance in the facility and all point sources are required to apply for and maintain these pollution permits. Theoretically, permissible levels are defined to meet the ambient air quality standards and since provincial authorities are responsible for meeting the national ambient air quality standards (which are specified by a ministerial decree), it is their environmental protection departments which set the allowable emission and discharge levels for each source.

To apply for a permit, each enterprise is required to conduct air dispersion modelling to determine the contribution of its facility's emissions to ambient air quality. The model results are then reviewed and approved by an independent expert before the provincial authority decides whether or not to issue the permit. Considerable importance is attached to the results of the air dispersion modelling, particularly when the applicant's emissions would contribute to violations of the ambient standard.

Polluters who have a valid permit must pay air pollution emission charges, including the SO_2 emission charge. Those polluters whose emission exceeds the emission standards specified in their pollution permits must pay fines. Generally, the fine imposed is based on the difference between the permitted and the actual level of emission. The fine rate to be paid when emissions exceed permitted levels by more than 1kg is 10 times higher than the regular emission charge rate.

An important exception to this process is the treatment of larger combustion sources. In 1990, a ministerial decree introduced special, technology-based emission standards for SO_2, nitrous oxide, and particulates, for major combustion sources with capacities greater than 200 kilowatts. Permitted emission levels for all of these large combustion sources were due to be determined according to the technological standards that were put in place in January 1998.

IMPLEMENTATION OF THE EMISSION CHARGE

All point sources which are required to obtain pollution permits for their operation are also liable for SO_2 charge payment. The charge payment is calculated on the basis of self-reported SO_2 emission. Provincial inspectorates are responsible for monitoring compliance with facility permits and verifying the accuracy of the pollution levels reported by the facilities.

In 1996, the SO_2 emission charge rate was US$94/ton (or PLZ2400 per ton).[4] Some sectors benefit from preferential rates or exemptions. For example, the charge rate for medicine manufacturers is ten times lower than the regular rate, and health and social care institutions, as well as educational and cultural organizations and prison management bodies, are completely exempt from the charge payment.

Table 11.1 shows the evolution of SO_2 emission charge levels over the last seven years. The most dramatic increases occurred in July 1990 and January 1991 when the charge system was revised. The table shows a clearly increasing trend for the nominal rate, though when expressed in US$, the rate has fluctuated somewhat. The 1992 rates quoted in the table were set in January and, later that same year, the new incoming government reduced all rates for emission charges, including the SO_2 rate, as a political response to strong protests from heavy industry against high charge rates. Surprisingly, these drastically lower charge rates also provoked strong protests from industry and the public. Those polluters who had already started abatement measures complained and revenues from pollution charges to the country's environmental funds dropped to about 32 per cent below the planned level. On April 1, 1993, as a result of bitter criticism coming from the press and public debates, new rates revoked most of the changes and all rates were reestablished at the previous levels.

Table 11.1 *Unit SO$_2$ charge rates*

	1990	July 1990	1991	1992	1993	1994	1995	1996	1997	1998
USD/ton	1.9	28.4	64.3	80.7	66.1	66.0	78.5	88.91	85.4	85.0
thousand PLZ/ton	18	270	680	1100	1200	1500	1900	2400	2800	3000

Source: Sleszynski (1996)

An annual adjustment procedure is a regular element of the implementation of the charge. Slippage in charge rates due to inflation is largely avoided by adjusting nominal charge rates to account for predicted inflation in the next year. If actual inflation deviates significantly in either direction from predicted inflation, adjustments can be made in the following year.

Environmental charges are collected once a year by the provincial administration environmental protection departments, though a 1990 amendment to the general legal act allows provincial governors to request quarterly instalments from large enterprises.

Enterprises are able to treat environmental charges as normal business expenses and to deduct the amount of charges paid from their taxable income. Charges are thus treated as a normal production cost. An interesting provision allows enterprises to deduct the amount of the charges *levied* in the current year from the current year taxable income even if they are delinquent in making payments and don't actually *pay* the charges until the next calendar year.

Annual revenue from collected SO$_2$ emission charges is relatively high in comparison to revenues from other environmental charges. Table 11.2 shows the revenue collected between 1990–1994.

Table 11.2 *Annual revenue from SO$_2$ pollution charge*

Year	Payment in million PLZ	Payment in million US$
1990	66,987	7.05
1991	1,283,796	121.31
1992	1,519,413	111.47
1993	3,376,827	186.10
1994	2,391,875	105.24
1995	3,232,028	133.47
1996	3,424,657	126.67

Source: Sleszynski (1996)

The revenue level varies year by year as a result of charge rate increases and changes in the effectiveness of collection. The high level in 1993, more than US$186 million, was largely caused by a delay in the payment of 1992 charges, due to the uncertainties surrounding the charge set that year.

The revenue collected is specifically earmarked for expenditure on air pollution control measures and is divided between three environmental funds in the following way:

- National Fund for Environmental Protection and
  Water Management (NF)   36%
- Regional Environmental Funds of the effected region[5]   54%
- Municipal Environmental Fund of the effected settlement[6]   10%

The revenue raised supports the operation of different subsidy schemes offered by the environmental funds.

In theory, SO_2 emission without a valid permit is penalized with a fine, equivalent to double the regular charge. In practice though, such payments are rare. Many polluters still operate without an emission permit. Some of the older enterprises can not obtain permits since they are unable to meet the obligatory emission standards, and many other polluters operate without permits because provincial authorities have limited resources to process or revise their permit applications. This unfortunate situation leads to discretionary decisions and bargaining with polluters.

DESIGN AND IMPLEMENTATION ISSUES

The Polish sulphur dioxide emission charge is rather unique because its rate is the highest in the world. The *relative* differences in the rates charged for different pollutants are correlated with differences in the pollutants' toxicities or their potential for causing environmental damages. For instance, air emissions of suspected carcinogenic pollutants such as benzene are assigned extremely high rates. The *absolute* magnitudes of the rates, however, are not set to reflect marginal damages to health and the environment, nor do they correspond to the marginal costs of abatement. While factors such as damage and abatement costs, and economic characteristics of the polluting sector, *are* taken into consideration in the setting of the rates, the charge levels are ultimately determined by what is considered politically acceptable and what is required to meet revenue requirements. Therefore, the charge is not designed to drive SO_2 emissions to the socially optimal level.

The question is then whether the charge still has the efficiency property – in other words, whether the charge induces abatement strategies by polluters, which would minimize the overall costs of pollution reduction. It is clear that the institutional context in Poland makes it unlikely that the conditions for efficient operation are met. The selection of cost-minimizing abatement strategies is feasible only if polluters can choose among different technologically and legally possible options. This is probably not the case for the Polish SO_2 charge. Polluters have no legal choice for exceeding their emission limit, as emphasized by the fact that the non-compliance fine is some ten times higher than the regular charge. Meanwhile, emission limits are rather strictly assessed, so only a few very expensive technological options are available for exceeding their emission limit. Certainly, the SO_2 charge rate is too low to provide an incentive to employ such expensive technology.

The SO_2 charge therefore collapses into a revenue-raising instrument, even with this relatively high rate. That does not mean, of course, that the charge has had no incentive effects at all. Some polluters, mostly the larger ones, *have* changed technologies and installed abatement measures. Many polluters, however, have chosen non-compliance and have avoided charge payments altogether.

The most common way of avoiding payment is simply to operate without a valid permit. In 1992, there were 46,305 facilities registered as air polluters. Although they are all required to apply for facility permits, it is estimated that nearly half of these facilities operate without valid permits. The backlog is largely attributable to limited local resources to process permit applications. Furthermore, many of the permits which do exist now need to be revised. When added to the resource and staff costs to prepare permits for those facilities currently operating without a permit, local environmental authorities will face an enormous challenge in trying to deal with this backlog.

Another potential source of non-compliance arises from the fact that the validity of self-reported emission levels is checked for only large polluters. The ability of the provincial inspectorates to verify self-reported emissions is limited by staff resources and compounded by the large number of facilities and individual pollutants that must be checked. Polluters are aware of these limitations and their reports probably understate their actual emission levels.

Due to the enforcement problems with the permits and the limited scope for verifying self-reported emissions, the amount of charges actually levied is well below the potential amount that could be imposed. Interestingly, despite these problems, the compliance rate for payment obligations from the levied charges is high, as shown in Table 11.3.

Table 11.3 *Charge Payment Compliance Rate*

Year	Compliance rate
1990	99%
1991	86%
1992	86%
1993	167%
1994	97%
1995	97%
1996	89%

Source: Sleszynski (1996)

The large annual variations in the compliance rate are mostly due to the political turmoil and uncertainties surrounding the level of charge rate in 1992. Many enterprises delayed their 1992 payments until the rates were firmly reestablished in 1993. The large volume of delayed payment made in 1993 raised the compliance rate to well over 100 per cent that year.

In comparison to other charges, the collection rate for the SO_2 charge is extremely high. The success may be attributed to the relatively effective monitoring and enforcement practice of larger sources with permits. This, in turn, is the result of a substantial effort to concentrate environmental protection policy on the problem of air pollution, and acid rain in particular. This makes the SO_2 charge one of the most effective revenue-raising sources in Poland.

The overall Polish environmental charge system is basically the only one in the region that has tackled the inflation problem successfully. The automatic annual revision of the charge rate is a legal requirement. Back in 1989, environmentalists managed to introduce legislation pegging the charge rates to the official inflation index. Unfortunately, the proponents of this measure did not realize that charges were pegged to the *ex post* inflation index, which only becomes available by the middle of the year after the one in which the charges have been applied. Therefore, in 1992 they switched to the present system where the *expected* inflation rate is used for adjustment.

Concerns about the low incentive effect and limited efficiency property of the SO_2 charge have prompted experiments with alternative instruments. Marketable discharge permits have been tested in Chorzów city and have received much attention in a comprehensive research study. In the experimental case, regional environmental inspectorates managed a transaction among two polluters exchanging SO_2 emission rights. The process of designing and implementing the transaction has made it clear that the regulatory and institutional framework of SO_2 emission control could not support a genuine, larger scale system. A marketable discharge permit approach is still not acceptable under the present environmental law, which is oriented towards regulating emissions from *individual* sources.

There is another ongoing open discussion on the introduction of an alternative instrument – product charges – motivated by the desire to maintain a revenue stream into the environmental funds system. The introduction of product charges, and particularly a differentiated fuel surcharge, will be critical in fulfilling the political desire to maintain substantial funding for the National Environmental Fund, while meeting the widespread expectation that the responsibility for the emission charges and the revenue from the non-compliance fines be retained at the provincial or municipal level. The most recent analysis on a possible fuel charge recommended the introduction of a 6 per cent surcharge for high sulphur content coal and high lead content gasoline, with no surcharge on the least noxious fuels, and intermediate rates of between 0 and 6 per cent for other fuels. At present, it is still not clear whether such a fuel surcharge will be introduced in the near future, or whether this introduction will be followed by specific changes in the operation of the existing SO_2 emission charge.

ENVIRONMENTAL AND ECONOMIC IMPACTS

It is impossible to attribute distinct and quantifiable environmental improvements to the operation of the emission charge. As in many other countries in transition, SO_2 emission rates decreased significantly during the course of a recession period. It is obvious that part of the SO_2 reduction has been caused by an economic decline at the early stage of the transition process. But later, in spite of a very ambitious economic growth of more than 5 per cent, SO_2 emissions have not increased at all, as shown in Table 11.4.

Table 11.4 SO_2 Emission in Poland

	1988	1989	1990	1991	1992	1993	1994	1995
SO_2 emission (million ton)	4.2	3.9	3.2	3.0	2.8	2.7	2.6	2.3

Source: Sleszynski (1996)

Table 11.5 shows the SO_2 emission intensity, calculated as SO_2 emission/GDP, and illustrates that emission reduction has been taking place despite the dynamic growth of the economy.

Table 11.5 SO_2 Emission intensity

	1988	1989	1990	1991	1992	1993	1994	1995
kg/million PLZ	6.61	6.17	5.73	5.75	5.27	4.91	4.46	3.74

Source: Sleszynski (1996)

According to OECD estimates, approximately 70 per cent of all violators in Poland are now complying with permit requirements. The 'list of 80' worst polluters and similar provincial lists based on the same concept, which include nearly 800 enterprises, have existed for almost five years. Public pressure and scrutiny on the actions of enterprises have led to considerable reductions in the amounts of pollutants emitted by the 'list of 80' enterprises.

Linking the decline in SO_2 emissions to a reduction in environmental damage is rather difficult. The environmental damage in question is a marked deterioration in the ecological quality of forests in the southwestern part of Poland. However, it is obvious that years of environmental neglect and pollution will continue to do damage for several decades, as the buffering abilities of the natural ecosystems have been eroded, making almost 50 per cent of Polish forests more sensitive and less sustainable. In spite of a positive trend in SO_2 emission, it will take something like 10–15 years before analysts can observe the first well-documented quality improvements in these seriously affected ecosystems.

The economic impact of the emission charges has been assessed in a sample of 112 of the largest polluters in Poland.[7] It has been estimated that pollution charges in 1992 represented 4.9 per cent of the total expenditures of these enterprises, abatement costs were 1.6 per cent and other pollution costs were 0.5 per cent. Total expenditures on charges for the surveyed enterprises amounted to US$190 million, compared to environmental fund investments of only US$108 million. This represents a heavy financial burden for a group of enterprises in transition. The energy sector, which has been the most affected by the emission charge, has experienced very diverse effects. The oldest, least-modernized power plants are still in trouble and can only respond to the policy by not complying or by postponing their payments of charges and fines, thereby creating a very difficult problem. Paradoxically, there are also power plants such as one in Jaworzno that have claimed that the SO_2 emission charge is too low to induce its internal technological restructuring.

An evaluation study of the environmental charge system looked at the distributional impact of the charges. Unfortunately, the analysis did not separate out the different charges, but the overall assessment is that the private sector pays a higher share in charges and receives a lower share in subsidies, in comparison to the public sector.

CONCLUSIONS

In the economic and institutional context of Poland's centrally planned economy, under which the implementation of the emission charge began, the charge could not play much of an incentive role. The soft budget constraint for state-owned enterprises meant that the extra financial burden of an environmental charge was more or less covered by additional money from the state. An interesting consequence of this institutional factor was that enterprises did not object too much to the introduction and collection of the pollution charges. In the period of an emerging market economy, the costs and profit implications of the SO_2 charge became real and it required political strength to reestablish the charge level.

The role of the charge in the present institutional context is a mixture of an incentive, an enforcement, and a revenue-raising tool. The full direct incentive impact of the SO_2 charge, even though it is relatively high, is greatly restricted by the strict individual emission limit system, under which the charge operates. It is an effective revenue-raising instrument, the revenue from which is earmarked for investment in curbing

SO_2 pollution. Since SO_2 emission is still a hot spot problem in Poland, the efforts of a powerful lobby of industrial polluters and a slightly less influential green lobby to concentrate the SO_2 revenue on emission reduction efforts seem justified. But the economic efficiency of environmental investments requires selection rules that are not subject to too much earmarking or any other discretional decision-making.

Given the practice of earmarking and project selection that may not achieve the greatest environmental benefits, it is interesting to look more closely at the largest industrial polluters and to speculate on whether they are better served by the current system of charges and recirculated grants and loans, compared to an alternative structure where charges are directly invested in environmental controls.

Several studies have suggested that more strict emission standards could be more effectively applied in the energy sector until the year 2004 by introducing a system of marketable discharge permits rather than by applying a progressive emission charge. This is particularly true for those power plants that are most affected economically by the charge payments and those which already have critical financial problems.

NOTES

1   Budapest University of Economics, Department of Microeconomics.
2   Warsaw University, Economics Department, Warsaw Ecological Economics Centre.
3   The term region here refers to what is sometimes called 'voidvodship', or 'wojewodztwo' in Polish.
4   PLZ = old Polish Zloty; PLZ 10,000 = PLN 1 (new Polish Zloty); in 1996, approximately PLZ 27,000 = US$1.
5   There are 49 regional funds.
6   There are 2,468 municipal funds.
7   The study was conducted by the Technical University of Bialystok.

REFERENCES

Directive of the Council of Ministers on Charges on the Use and Alteration of Natural Environment, DzU No 133, pos 638, December 27, 1993; amendments: Dz U No 51, pos 203, and Dz U No 140, pos 772 of 1994; Dz U No 153, pos 775, October 19, 1995.

OECD (1995) *Environmental Performance Review of Poland*. Centre for Cooperation with Economies in Transition, OECD, Paris.

Sleszynski, J (1996) Two Case Studies on Implementation of Selected Economic Instruments in Poland. Paper submitted in the UNEP project on Economic Instruments, Warsaw.

Zylicz, T (1993) Case Study on Poland. Paper presented at the OECD Workshop on Taxation and the Environment in Economies in Transition, Paris.

12

PRODUCT CHARGE ON PACKAGING IN HUNGARY

Zsuzsa Balogh[1] and Zsuzsa Lehoczki[2]

BACKGROUND

The environmentally safe treatment of municipal solid waste has recently become a problem in Hungary. The estimated volume of municipal solid waste generated in 1995 is approximately 20–21 million m^3 (between 4 and 4.5 million tons), or 1.8 m^3 per capita. The trend is a 2–3 per cent annual increase in the volume generated. According to experts' estimates, about 70 per cent of this solid waste is disposed, about 8 per cent is incinerated, and only 2 per cent is recycled. There is no information on the remaining 20 per cent, which most likely ends up as illegal littering. At the same time, the proportion of reusable material in municipal solid waste is increasing. The available capacity of the existing disposal sites is estimated to accommodate about 60 million m^3 of additional solid waste.[3]

Packaging material constitutes around 20 per cent of the municipal solid waste. A major problem with packaging waste is its high volume and low density. It fills up disposal sites quickly and finding new sites is difficult and costly. There are, however, several options for mitigating the potentially negative environmental impacts of packaging. Different forms of recycling are possible and are already being practised in some countries, to reduce packaging waste. Encouraging the implementation of such options can increase the presently low ratio of recycled waste.

The role of packaging as a marketing tool was not so important before the transition to a market economy in 1989. Prior to the transition, there was an obligatory deposit refund system for different types of packaging, and in particular for glass containers. The transition resulted in an increase in packaging and the loss of the legal basis for the deposit-refund system. Regulations related to deposit-refunds were part of the centrally-set price system, and in the first phase of the transition most prices were liberalized and previous legislation on price setting was replaced. In 1992, when food-processing companies refused to recycle their preserving jars, it became apparent that the new act on prices had ignored the issues of the deposit-refund system. The need to devise and introduce policy instruments aimed at establishing the framework and motivation for making recycling an integral part of the packaging industry had now become clear. Responding to that need, the Ministry of Environment and Regional Policy drafted a proposal for introducing a charge on packaging material in February 1993. The proposal sparked off fierce opposition by all affected parties and it underwent several major transformations in the long negotiation process. Finally, the packaging charge chapter of Act LVI of 1995 on Environmental Product Charges came into force on January 1, 1996.[4]

THE PRODUCT CHARGE

The environmental charge on packaging is a product charge, with a structure very similar to an excise tax. The act on product charges has established the general aims of any product charges applied in the Hungarian environmental policy: (i) to raise the financial resources needed to mitigate the environmental damage caused by a product which is hazardous to the environment at any point of its lifecycle; and (ii) to provide an incentive to reduce this environmental hazard.

This general objective implies that the product charge on packaging was also introduced as a fiscal instrument to raise revenue earmarked for environmental protection measures related to packaging. The incentive impact can be expected from the differentiated unit charge rate, though the rates have not been fully established to reflect differences in environmental damages. Another incentive impact is intended by the application of discounted charge rates for products with eco-labels or exemptions, which can be acquired by complying with obligatory recycling rates.

According to the general structure of the product charges, a charge must be paid on the defined products at the first point of sale or internal use for domestically produced goods, or at the point of levying custom duty, for imported goods. It is possible to halve the unit charge rate for any chargeable product, by acquiring an environmentally friendly label (eco-label). In the case of packaging, full or partial exemptions can also be acquired by meeting obligatory recycling targets, fully or partially. In the interest of international competitiveness, export products are exempt from packaging product charges.

Product charges on domestically produced products are collected by the Tax Collection Office, while charge liabilities on imported goods are collected by the Custom Office. The packaging product charge revenue, as with other product charge revenue, is transferred to an extra-budgetary fund called the Central Environmental Protection Fund (CEPF).

The product charge revenue is spent through the CEPF. According to the Fund's spending regulations, a part of the income can be used for financing public tasks in the field of environmental protection, but at least 85 per cent of the revenue must be spent on subsidizing environmental protection projects. Subsidies can be given for financing both investment and recurrent costs.

IMPLEMENTATION OF THE PACKAGING PRODUCT CHARGE

The specific legislation on the packaging product charge establishes payment obligations for those persons who first sell or import packaging material, either as a separate product or as part of another product. Payment is based on the weight of the packaging material, according to the following formula.

PAYMENT = (1–Scale of Exemption) * Weight of Packaging * Unit Charge

The unit charge is defined for the unit weight of the packaging (HUF/kg)[5] and differentiated according to the material (see Table 12.1).

Table 12.1 *Unit Charge Rates.*

Material of Packaging	Product Charge (HUF/kg)		Product Charge (US$/kg)	
	1996	*1998*	*1996*	*1998*
Plastic	10	11	0.070	0.054
Mixed	8	9	0.056	0.044
Aluminium	5	5.4	0.355	0.026
Other metals	4	4.3	0.028	0.021
Paper, wood, natural textile	3	3.3	0.021	0.016
Glass	2	2.2	0.014	0.011
Other	5	5.4	0.035	0.026

The recycling ratio used for determining eligibility for partial or full exemptions from the payment was defined in a ministerial decree, as the proportion of packaging material collected and reutilized or incinerated for energy generation. Table 12.2 shows the recycling ratio in operation up until 1997. The Ministry of Environment and Regional Policy plans to increase it, but that is still being debated by the affected parties.

Exemptions can be gained, depending on the ratio between the actual and obligatory recycling. The scale of exemption is calculated on the basis of the following formula, where the value of S can vary between 0 and 1.

Table 12.2 *Obligatory Recycling Ratio*

Packaging Material	Obligatory Treatment Ratio (Reutilization or Energy Generation) (%)		
	1996	1997	1998
Paper	30.5	45.0	45.0
Glass	20.0	35.0	35.0
Other Material	12.5	25.0	25.0

$$S = [(R+E) / (T * a)]$$

where:

S = scale of exemption;
R = weight of recycled packaging;
E = weight of recollected packaging used for energy generation;
T = total weight of packaging (base for product charge payment);
a = obligatory recycling ratio (as stated in Table 12.2).

Exemptions can be gained if the potential charge payers recycle the packaging waste themselves or establish a legally-binding contract with an authorized recycling company. The contracting option has been introduced to encourage the establishment of a recycling system similar to the one operating in Germany.

Those persons whose activities require them to pay the charge must submit an initial report on a standard form providing basic payment-related information. Actual payment must be calculated and paid monthly. Payment is based on self-reporting, and supporting documents have to be presented to the collecting organizations on request. Quarterly reports must be prepared with detailed information to justify the payment calculations for each month in the given quarter. There is a standard sheet to be filled in for these reports. Payment must be made to the special bank account of the Tax Collection Office or the Custom Office. Money accumulated on the special accounts is transferred to the Central Environmental Protection Fund on a monthly basis.

Requests for exemptions must be submitted to the Regional Environmental Inspectorates, who then attach their comments and forward the applications to the Central Environmental Fund Management Unit. The evaluated requests are presented to the Ministry of Environment and Regional Policy, where final decisions are made. Environmentally friendly labelling of a product can be acquired through a standard evaluation process, managed by a public company. Requests for export reimbursement must be made to the Central Environmental Fund Management Unit, where decisions are made in liaison with ministerial experts.

The product charge legislation supported the establishment of a consulting body and special forum called the Packaging Materials Environmental Product Charge Technical Committee. The Committee is responsible for the following:

- proposing the required proportion of treatment of packaging wastes;
- developing recommendations for rules on the selective collection of packaging materials, for the local governments that voluntarily join the selective collection;
- reviewing licences for the establishment of coordinating bodies for recycling schemes;
- proposing financial assistance objectives for the environmental fund; and
- taking part in awarding financial assistance.

Seven members of the Committee are delegated by professional associations, two by environmental NGOs, two by local government; and one each by the Ministry of the Interior, the Ministry of Agriculture, the Ministry of Industry and Trade, the Ministry of Transport, Telecommunications and Water Management, the Ministry of Welfare, the Ministry of Finance and the Ministry of Environment and Regional Policy. The representative of the Consumer Protection Agency is a permanent invitee to the committee meetings. The president of the Committee is appointed by the Minister of the Environment and Regional Policy. The Committee meets at least once every quarter, and more often if the need arises.

The major tasks of monitoring and controlling both the collection and disbursement of revenue rest with the Environmental Protection Fund's Management Unit. Reporting requirements are in line with the general rules of public budgeting. Annual reports must be prepared and presented to Parliament. According to the new state administration act, detailed data on the trends of revenues and expenditures must be presented in the annual state budget act. Special rules govern the publication of the annual spending programmes and the results of financial assistance decisions.

NEGOTIATIONS DURING THE PREPARATORY PROCESS

The packaging product charge was introduced after much preparatory work and lengthy negotiations. In February 1993, the first proposal was drawn up for a product charge on packing materials of liquid foodstuffs. There was great resistance on the part of the most affected economic actors. They complained that the proposal was not comprehensive and the charges were too differentiated. In April 1993, the Parliamentary Committee for the Environment put the proposal on its agenda, and – in spite of fierce opposition expressed in the Committee meeting – it was proposed that the product charge be amended and extended to all packing materials. An amended proposal was prepared and negotiations continued.

At the end of 1993, negotiations ended with an agreement between the Ministry of Environment and Regional Policy and several trade organizations representing the most affected companies, including many multinational corporations. The agreement was criticized by NGOs for giving too much emphasis to the views and interests of industry. In 1994, various revised versions of the proposed legislation were circulated and discussed with reference to the already-signed agreement. By the beginning of 1995, the newly-elected leadership of the Ministry decided to strike a compromise with the trade organizations, which were now prepared to be constructive since they wanted to maintain their image of being concerned about the environment.

A further round of negotiations resulted in a new agreement, in which the trade organizations agreed to the extended scope of the charge. The Ministry, at the same time, agreed to a long list of possibilities for waiving or reimbursing charge payments, and to reconsider the product charge in the context of a comprehensive solid waste management legislation. The revised agreement was harshly criticized by NGOs again but they allied themselves with the trade organizations to get the revenue earmarked for the financing of recycling and reuse of packaging material.

IMPACTS

The charge is still too new to discuss extensively its ultimate environmental impact. Packaging waste recycling measures, however, have already been initiated. The most affected packaging material producers and users (such as Tetra Pack Co and Coca Cola Co) took steps to organize the collection and reuse of bottles and other forms of packaging. These efforts have been largely reinforced and hastened by the present legislation with its extended set of exemptions for those companies who meet recycling targets.

The unit charge rates were determined through the same process of political negotiations. The rates, however, are ultimately based on the revenue-raising objective of the charge. This has required the calculation of the costs of treating packaging waste. These waste treatment costs depend on several factors:

- the type of products inside the packaging: if the packaging covers hazardous products, the packing material itself becomes hazardous waste, thereby increasing the treatment costs;
- the type of treatment of the packaging: whether it is recycled (multi-way bottles), used for energy generation (burning polyethylene bottles) or deposited in waste yards.

There have been several attempts to use a lifecycle analysis to determine the environmental damage caused by different types of packaging material but doubts have been raised as to the certainty of the results produced by such a (rather expensive) analysis.

The present rates also reflect concerns about how much the charge would add to the price of basic food items such as milk. The unit charge amounts to between 1 and 10 per cent of the price of the packing materials. Responding to the concerns about equity, the product charge payment remains below 1 per cent of the price of food packed, such as milk or other dairy products. The inflationary impact of the charge was estimated to be low, although the methodology used for the assessment was rather unrefined.

The charge system in its present form is quite comprehensive and entails a complex payment collection and administration structure. First of all, there are a large number of potential charge payers. The reason for levying the product charge at the first point of sale or at the point of import is to reduce the number of charge payers. But in the case of packaging material this does not work very well, since materials can be used for many other purposes apart from packaging. Plastic foil for example can become liable for charge payment when it is used to wrap cheese in a small grocery shop, so thousands of small shops are potential charge payers. Meanwhile, large producers of foil can claim that when they sell their product it is not certain that it will be used for packaging.

There were around 700 charge payers in 1996, compared to under 20 for any other product charge in Hungary. The average monthly payment of a charge payer has been around US$1000. In 1996, charge payments were collected by the environmental fund administration, which has a limited capacity for controlling self-reporting and for enforcing payment. Since 1997, collection responsibilities have shifted to the tax authorities, in the hope that it would increase collection efficiency. The Tax Collection Office – similarly to the Custom Collection Office – charges around 3.5 per cent fee for collection and an additional 1 per cent for administering exemptions and export rebates.

The planned packaging product charge revenue figure for 1996 is 4,000 million HUF, according to the official government budget act for 1996. Actual revenue figures for 1996, plus the expected and planned figures for 1997 and 1998, respectively, are presented in Table 12.3.

Table 12.3 *Charge Payment in 1996*

	million HUF	*million US$*
1996 planned charge payment	4000	28.1
1996 actual charge payment	1924	12.6
1997 expected charge payment	3287	17.6
1998 planned charge payment	3650	17.8

Source: Government budget proposal, 1996; calculation based on personal interviews

Slightly more than half of the payment (52 per cent) comes from charges levied on domestically produced packaging material and the rest is from imported packaging. It is clear from Table 12.3 that the actual charge payment in 1996 is less than half of the planned figure, indicating a large overestimation in the budgetary revenue expectations. In 1997, revenue rose sharply. It is likely that this was the result of improved collection efficiency, due to the shift of collection responsibilities to the Tax Collection Office.

This gap can be partly explained by problems in enforcing the charge, but is mostly due to exemptions. Companies were allowed to submit applications for exemptions prior to the packaging product charge coming into effect. Many of the large packaging material users and traders were well-prepared and seized the opportunity and have not paid any product charges in 1996. Broad estimates of the revenue loss due to exemptions range from 10 to 30 per cent, even though exemption permits are given to only 2–5 per cent of the charge payers.

The impact of lower charge payments for environmentally friendly labelling has been negligible, due to the fact that detailed assessment guidelines have been developed for only a few packaging materials, so this option has not been widely available.

A major environmental impact is expected from the spending of packaging product charge revenue. Originally, the entire revenue was earmarked for spending on subsidies for packaging-related environmental measures. The main target for spending is the subsidizing of packaging recycling infrastructure and operation. Similar efforts in Germany, Austria and France have been studied in setting up the packaging charge system in Hungary. However, the drafting of the annual spending programme has been somewhat slow and so no actual disbursements took place in 1996.

CONCLUSIONS

The product charge on packing materials is a temporary regulation, which will be replaced by an appropriate section in the forthcoming acts on solid waste management and packing material waste. These two acts are envisioned to provide an integrated set of obligations for all the actors in the packing chain. Draft versions of each of these acts are under discussion but it is unlikely that they will be passed in the near future.

In the long run, the objective of a product charge on packing materials is to create the conditions to correspond to the EU packing directive issued in December 1994. It is important for harmonization that recycling and treatment of packing materials take place on an increased scale and that packing materials are tracked both qualitatively and quantitatively. There are plans for the new packaging waste act to switch from being a product charge to a set of legal obligations for recycling and treatment, as the necessary recycling infrastructure would be built by then with the financial support of the environmental fund.

The main lesson from the preparation and negotiations of the product charges is that the government must negotiate with powerful affected parties, who are able to clearly articulate their own interests. Environmental policy makers have to be prepared to formulate their proposals in a fairly complicated process, which is sometimes dominated by strong economic pressure groups. Experience also shows that negotiations are often based *not* on sound professional assessments, but on the political appeal of certain options.

The broad range of packaging materials covered in the charge scheme, and the extended possibilities for requesting exemptions, make the system difficult and costly to enforce. However, once the necessary administrative framework is implemented, the relative cost declines. Moreover, the charge has had a substantial environmental impact even in the preparatory phase. When strong signals were sent that packaging waste was high on the environmental policy agenda, the larger companies started their own preventive and recycling measures. They could not afford to ignore the public relations and marketing impact of being perceived as not caring for the environment.

NOTES

1   EXTERNAL Environmental Economics Consulting Ltd.

2   Budapest University of Economics, Department of Microeconomics.

3   Nemzeti Környezetvédelmi Program,1996.

4   The act introduced product charge on batteries, refrigerators/refrigerants and tyres, as well as packaging material, and amended the existing legislation on the transport fuel charge.

5   HUF=Hungarian Forint; in 1996 approximately HUF153 = US$1.

REFERENCES

Balogh, Z (1996) A környezetvédelmi termékdíjak bevezetésének gyakorlati kérdésdései (Some practical questions on the introduction of environmental product charges). Budapest.

Barta, J (1994) Háttértanulmány a csomagolóeszközök környezetvédelmi termékdíjának bevezetéséhez (Background study to the introduction of the packaging product charge). Ministry of Environment and Regional Policy, mimeo.

Government budget proposal (1996)

Ministry of Environment and Regional Policy (1996) *Guidelines for the Environmental Product Charges.*

Ministry of Environment and Regional Policy (1996) Nemzeti Környezetvédelmi Program (National Environmental Protection Program).

Ministry of Environment and Regional Policy (1996) Támogatási irányelvek, tervezetek (Spending Programme for the Environmental Product Charges).

13

PRODUCT CHARGE ON
TRANSPORT FUEL IN HUNGARY

Zsuzsa Lehoczki[1]

BACKGROUND

Hungary's transition from a centrally-planned economy to a market economy induced major structural changes and resulted in serious industrial decline, which in turn led to improvements in air quality in the early 1990s. According to a compound measure of air pollution, 38 per cent of the population lived in polluted areas in 1983–1984 and this had declined to 28 per cent by 1993. This general trend masks important differences for different pollutants. In 1993, only 1.6 per cent of the population was regularly exposed to sulphur dioxide (SO_2) concentrations higher than the ambient standard, while some 4.2 per cent were exposed to excessive concentrations of particulate matter, and 25.7 per cent to excess levels of nitrous oxide (NO_x).[2] The much higher exposure to high NO_x levels can be explained by differences in the emission trends of different sectors. Industrial emissions have declined since 1980 with a particularly sharp decline in 1990–1991, due to the severe recession. Air pollution from the power sector has also declined steadily. The only sector that has not shown a considerable decline in pollution levels is transport, which is the major source of NO_x emissions. Table 13.1 indicates that in addition to NO_x, more than half of the carbon monoxide (CO) and volatile organic compounds (VOCs) emissions also originate from the transport sector.

The high exposure rate for NO_x is also due to the concentration pattern of emissions. Transport-related pollution problems are concentrated along major highways and in big cities, particularly in Budapest where 20 per cent of the country's population live. In 1991, ambient air quality levels in Budapest often exceeded the Hungarian Air Quality Standards for nitrogen oxides, lead, hydrocarbons, soot and sulphur

Table 13.1 *Transport Sector Emissions (kt/yr) and the Sector's Share in Total Emissions*

	NO_x		SO_2		CO		VOCs	
	Emission	Share	Emission	Share	Emission	Share	Emission	Share
1980	111	40%	49	3.0%	na	na	na	na
1985	110	41%	21	1.5%	na	na	na	na
1989	117	47%	16	1.5%	na	na	91	44%
1991	98	49%	14	1.5%	487	53%	73	51%
1992	94	51%	13	1.6%	491	59%	68	50%
1993	92	50%	8	1.0%	452	57%	73	51%
1994	94	50%	7	1.0%	437	57%	71	50%
1995	101	53%	7.5	1.1%	449	59%	73	49%

Note: na=data not available
Source: National Environmental Protection Programme, KTM, 1994; EGI, 1997

dioxide, and occasionally for carbon monoxide too.[3] Transport is the dominant source of all of these emissions except SO_2, contributing 57 per cent of NO_x, 80 per cent of lead, 81 per cent of CO and 75 per cent of hydrocarbons emissions.

There are a number of environmental problems related to emissions from transport but the most significant one is the risk to human health. Several studies have been conducted to assess transport-related environmental damage and some of these have expressed the damage in monetary terms. The annual damage estimates in 1991 range between HUF8800 million and HUF69,600 million for the country as a whole, and between HUF4440 million and HUF34,800 million for Budapest.[4] The large variations in estimates reflect large data and methodological uncertainties in the studies. Considering that even the lowest estimate of the national damage amounts to 0.7 per cent of GDP, particular attention to reduce pollution from the transport sector is well justified.

To address these environmental problems, a programme has been developed jointly by the Ministry of the Environment and Regional Policy and the Ministry of Transport, Telecommunication and Water Management, in collaboration with leading experts. An environmental product charge on transport fuel was introduced in 1992 to provide financial backing for the implementation of this programme.

THE TRANSPORT FUEL CHARGE

The environmental product charge on transport fuel was the first economic instrument in Hungary which aimed to tax the consumption of a polluting product. It is an input product charge, with a structure very similar to that of an excise tax. The revenue-raising aspect of the charge has been particularly pronounced, as it was introduced explicitly to raise funds to finance a transport-related environmental protection programme.

Transport fuels cause air pollution while they are being consumed in moving vehicles. Emission levels from transport depend on various factors, including: the quality of the input fuel, several technical parameters, the age of a vehicle fleet, and the density and organization of traffic. Reductions in emissions can be achieved through interventions at any of these points, with varying effectiveness and costs. Applying a direct emission charge can be economically efficient and environmentally effective and can result in the reduction of emissions via least costs measures affecting any of the above factors. Direct emission taxation, however, is not feasible, since the large number of sources would make controlling and enforcing such a tax highly expensive if not impossible. A charge imposed on fuel as an input to transport is a feasible, close substitute to a direct emission tax. It is closely correlated to the emission being regulated, and its administration costs are low.

In Hungary, a transport fuel charge must be paid at the first point of sale or internal use, for domestically produced goods, or at the point of levying custom duty, for imported goods. The basis for the product charge payment is the quantity of fuel distributed by the obligant within the domestic market, or the quantity of products used for their own internal business activities. In the case of marketing a product subject to a product charge, the charge has to be included in the net price of the product, before calculating the Value Added Tax (VAT).

No reductions or exemptions are available for the transport fuel product charge. A rebate on exported transport fuel is possible if the product charge has already been paid at the first point of domestic distribution. The rebate is given if the applicant can supply the customs declaration as proof that the product is being exported, and the relevant document proving prior payment of the product charge.

Revenues from the transport fuel charge are channelled into the Central Environmental Protection Fund (CEPF) and must be spent according to the CEPF's spending regulations. Part of the income can be used for financing public tasks in the field of environmental protection, including the costs of charge collection and monitoring and enforcement expenditures. At least 75 per cent of the revenue must be spent on subsidizing environmental protection projects. Subsidies can be given for financing both investment and recurrent costs. Disbursements can be made in the form of grants, soft loans, interest subsidies, or loan guarantees.

IMPLEMENTATION OF THE TRANSPORT FUEL CHARGE

The transport fuel charge was introduced by an act which came into effect in 1992. This act was subsequently replaced by a comprehensive

Act on Environmental Product Charges in 1995. The new act stipulates the general features of any product charge, and contains sections on specific product charges. In addition to the transport fuel charge, four other product charges were introduced by the act.[5]

The administration and disbursement rules of the charge have changed somewhat since its introduction. A major change took place in 1996 when collection responsibilities shifted from the CEPF administration to the Custom Office in the case of imported transport fuel, and to the Tax Collection Office in the case of domestically-produced fuel.

The unit charge rate has been increased several times. The higher rates were justified by the need to protect its real value and the need to increase available resources for transport-related and general air pollution reduction measures. The unit rate has not been changed since 1995 but new product charges were introduced on other fuel products (for example, heating oil). Table 13.2 shows the different rates in effect between 1992 and 1997.

Table 13.2 *Unit Charge for Environmental Transport Fuel Charge, 1992–1997*

| | Unit Charge Rate | | | | |
	May 1992– Dec 1993	Jan 1994– Oct 1994	Nov 1994– Dec 1994	Jan 1995– Sep 1995	Oct 1995– 1997
US$/l	0.006	0.009	0.010	0.011	0.015
(HUF/l)	(0.5)	(0.8)	(1.0)	(1.2)	(2.0)

Source: Government Budget Acts, 1994, 1995, 1996, 1997

Those persons whose activities create a charge payment obligation must submit an initial report on a standard form, providing information on their economic status, which is required to calculate their payment obligations. Total payment is calculated as the unit charge rate multiplied by the quantity of transport fuel. The quantity has two elements: the volume of imported transport fuel, and the quantity of domestically-produced transport fuel at its first point of commercial distribution. Payment must be made each month and is based on self reports and supporting documents presented to the collecting organizations on request. Quarterly reports need to be prepared, giving detailed information on payment calculations for each month in the given quarter.

Payment must be made to a special environmental product charge bank account of the Tax Collection Office and the Custom Office. The revenue accumulated in the special accounts is then transferred to the Central Environmental Protection Fund on a monthly basis. The revenue stream has grown steadily since its introduction, as shown in Table 13.3.

Table 13.3 *Revenues to the CEPF, 1992–1996*

	1992	1993	1994	1995	1996	1997*	1998**
Transport fuel product charge,							
in million US$	11.1	15.9	23.8	27.1	43.8	39.8	35.6
in million HUF	879	1460	2502	3400	6694	7431.3	7300
Total CEPF revenue,							
in million US$	20.7	29.3	57.8	43.8	93.1	85.3	116.4
in million HUF	1630	2698	6079	5505	14228	15930	23860

Notes: *planned; **expected
Source: Government budget proposals, 1993, 1994, 1995, 1996, 1997, 1998

The revenue goes to the Central Environmental Protection Fund. Originally, the charge revenue was earmarked strictly for subsidizing transport-related environmental measures. In 1994, however, the scope of eligible spending was extended and it became possible to support air pollution reduction programmes in the most heavily polluted cities. In 1995, only half of the charge revenue remained earmarked for the reduction of pollution from transport, and since 1997 all of the fuel charge has been untied and can be used for any spending from the Fund.

DESIGN AND IMPLEMENTATION ISSUES

While direct emission charges or fees have a long tradition in the Central Eastern European countries, the fuel product charge was a new instrument, not only in Hunagry but in the region as a whole. Drafting and negotiating the related legislation required a great deal of preparatory work. Preparation, however, concentrated almost entirely on designing a revenue spending programme since it was important to generate professional, political and public support for the product charge. The following analysis examines the direct economic and environmental impacts of the charge, as well as some aspects of the revenue spending.

The overall programme which has been developed to address transport-related problems is made up of the following components:

- Modifying the emission characteristics of the country's vehicle fleet, by changing the technical requirements and/or the age distribution of the vehicles;
- Modifying the emission potential of transport fuel;
- Infrastructure development and traffic management; and
- Collecting and recycling transport related waste.

Specific activities within these components were elaborated into a three-year programme to be financially assisted through the environmental fund. Elements of this programme include:[6]

- *Two stroke engine programme:* installing new engines; retrofitting catalytic converters; encouraging accelerated retiring of vehicles.
- *Bus programme:* installation of 'smoke reducers'; retrofitting 'turbo filters'; accelerated replacement of the oldest buses.
- *Four stroke engine programme:* retrofitting catalytic converters.
- *Alternative fuels:* support infrastructure for using natural gas as a transport fuel; supporting agricultural-based bio-fuel production.
- *Infrastructure – urban transport:* road construction and repaving at the most critical junctures; parking lot construction; improvements in traffic flow patterns; noise barrier construction; tree planting along busy roads; building bicycle paths; traffic reduction in residential areas.
- *Public transport:* acquisition of new vehicles; subsidies for mass transit fares.
- *Waste management:* reduction of damage from land disposal.

Although the environmental fund revenue is certainly not sufficient to finance major infrastructure development projects, it *is* able to enhance the impact of ongoing investments. This assumption has served as the basis for revenue target calculation and for establishing unit charge rate proposals.

Strict economic efficiency criteria were not applied in determining the proposed unit charge rate. Economic efficiency calls for an overall tax rate that is equal to the amount of the external damage caused by a unit of transport fuel. No attempt was made to relate all the tax-type payments imposed on transport fuel to the environmental damage caused by the consumption of this fuel.

The product charge on transport fuel was negotiated and introduced in the midst of a complete restructuring of oil products pricing and retail. Prior to 1991, the retail sale of oil products was in the hands of a state-owned company. Multinational oil companies had only a small percentage of the oil retail market and were represented by only a few gas stations. The retail price, as with most other prices, was centrally set and controlled. In 1990, prices were removed from state control, with a few exceptions. End use prices for oil products were among those prices which the state retained the right to set, with rules for following the world market price. Oil import and export were also under strict state control. The tax component of the oil price was very high and the government was accused of masking an increase in the tax component by not allowing the price of oil to fall while world market prices declined.

Social pressures forced the government to accelerate the relinquishing of their control of the gasoline market.[7] The dominant state-owned company, which owned the refineries and distribution network, was reorganized under a new name, MOL Ltd. In 1992, when liberalization of the oil product market took place, trade restrictions were lifted, duties on the import of oil products were abolished and government control over oil pricing was removed. Retail and wholesale prices for all oil products are now set by the market. MOL Ltd., which still supplies almost three-quarters of the Hungarian market, sets its ex-refinery prices in line with international market levels in order to compete with imported products. Lifting the trade restrictions has resulted in the rapid emergence of a large number of small private trading companies. In the first half of 1994, companies other than MOL accounted for 92 per cent of the total imports of oil products. Following liberalization of the oil product market, a number of foreign oil companies have established a presence in Hungary, giving rise to a large increase in the number of service stations and creating greater competition. MOL's market share fell to an estimated 32 per cent in 1993.[8]

The type and level of taxation on transport fuel was largely determined in debates which took place during 1991and 1992. The proposal on an environmental product charge became part of these negotiations. The proposed charge rate, however, was so small in comparison to other charges on transport fuel, including an excise tax and a road fund contribution, that it did not generate large public attention. Finally, a rate of 0.5 HUF/l (0.06 US$/l) – half of what was originally proposed – was accepted and formulated into the act of 1992.

The preparatory negotiations for the design of the charge provide an example of how the revenue-raising function of an environmental charge can contradict the requirements for an incentive impact. In 1992, leaded gasoline consumption was high and, because of this, the original product charge proposal included differentiated charge rates. The rate of 1 HUF/l (0.012 US$/l) was proposed for leaded gasoline and diesel fuel, and a lower rate of 0.8 HUF/l (0.01 US$/l) for unleaded gasoline. The differentiated product charge was part of a larger programme aimed at phasing out leaded gasoline. However, when the charge rate was reduced in the process of parliamentary discussions, it became obvious that maintaining these rate differentials would have resulted in a very low rate for unleaded gasoline. Even though differentiation would have set the stage for a sharp decline in the use of leaded gasoline, it would also have drastically reduced the revenue-raising potential of the fuel charge. In the end, a uniform rate was passed, accompanied by a commitment that assistance would be given to phase out lead through disbursements from the charge revenue.

A potential direct incentive impact of the charge would have been a decline in the use of transport fuel. Such an impact can hardly be expected, with so low a unit rate. Nevertheless, the demand for oil products did indeed fall substantially, by 6 per cent in 1991 and by a further 1 per cent in 1992. This decline can be attributed to two major factors: the recession, and a drastic increase in oil prices. The government has steadily increased excise duties and taxes on the sale of motor fuels, largely in response to the need for revenue. In mid 1994, taxes and duties represented about 70–75 per cent of the retail price of gasoline. The government has now established a differential between duties on leaded and unleaded gasoline, in line with EU practice to encourage the phasing out of leaded fuel. In addition to the product charge, a road fund tax, a stockpiling fee, and an excise duty of 25 per cent VAT are charged on transport fuels. Table 13.4 gives actual figures for these tax components in 1995.

Table 13.4 *Tax Components of the Transport Fuel Price in 1995 (in HUF)*

	Leaded	Unleaded	Diesel
Road use charge	9.50	9.50	9.50
Environmental product charge	1.20	1.20	1.20
Excise Duty	41.60	35.30	30.20
VAT		25% of the sales price	

A behavioural impact of the product charge is difficult to demonstrate, since the charge rate amounts to about 1.5 per cent of the market price of transport fuels and less than 3 per cent of the total tax component of that price. Considering the price inelasticity of the demand for transport fuel, such a small addition to the cost of transport fuel can have little impact on its consumption.

A major concern in the preparatory phase was how the real value of the charge could be protected against inflation. Indexing with some inflation indicator was not considered because Hungarian economic and monetary policy makers generally rejected such a practice in any area of public policy. The alternative structure proposed to address the problem of inflation was to define the charge rate as a percentage of the sales price. However the Ministry of Finance did not accept that option either. It was true that the full administrative consequences of such an option were not elaborated and would have probably been more serious than was outlined in the proposal. The lack of any special provision for maintaining the real value of the charge has not created problems, since the unit charge has been increased without much debate in connection with each government budget law and/or with amendments to product charge legislation.

The administration of the charge is fairly simple, due to the very limited number of refineries in the country (see Table 13.5) and the high costs of transporting gasoline, which greatly limit its import and export. A major part of the payment comes from gasoline from domestic refineries, which contribute approximately 80 per cent of the total transport fuel charge payment.

The payment compliance rate is basically 100 per cent, among those companies who report payment obligation. Non-compliance issues stem from evasion of the really biting consumption tax and VAT. As general tax payment enforcement improves, so will compliance with the product charge.

Table 13.5 *Some Characteristics of the Transport Fuel Product Charge Payment in 1996*

	Number of charge payers	Average monthly charge payment by each payer (US$/month/payer)	Ratio of export rebate over total charge payment
Transport fuel charge	18	159,000	0.17%
all five product charges	738	5180	0.36%

Source: Calculations based on personal interviews at the Ministry of Environment and Regional Policy

It is fairly easy to calculate the revenue generated by the gasoline charge. Transport fuel consumption is stable and annual revenue estimates are prepared in connection with the annual budget preparation for the Central Environmental Protection Fund. These estimates have proven quite accurate.

An assessment of the charge's environmental impact and economic effectiveness needs to incorporate an analysis of the disbursements of the revenue collected. A cost-effectiveness and cost-benefit analysis of the major programme elements was carried out as part of the charge preparation process.[9] The preparatory work focused on evaluating options for the expenditure programme, using cost-benefit and cost-effectiveness techniques. These analyses were requested in order to help design specific programmes that would be economically efficient, and to evaluate whether some of the programmes deserved greater levels of expenditure than others.

The analysis findings suggest that most cost effective measures relate to public transport. Improving emission characteristics and the age distribution of buses are effective and efficient ways of reducing urban air pollution. Replacing diesel bus engines and applying bus turbo chargers also have a high net benefit value. Retrofitting catalytic converters produces rather moderate levels of net benefit, while engine replacement in two stroke cars produces a negative benefit. The costs of infrastructure development such as Park and Ride investments are so high that they also yield a negative benefit, if one considers only the environmental benefits.

Actual spending from the environmental fund started rather slowly. In 1992, new spending rules were introduced, many of which aimed at facilitating programme-based spending. The environmental policy makers and the CEPF administration found it difficult to develop detailed programme and project appraisal guidelines, though they did manage to launch some programme elements and were reasonably successful in implementing the bus engine replacement programme. The programmes for retrofitting catalytic converters had a rather slow start and indeed it was impossible to generate high demand for such retro-fitting unless it was offered at a very low price. A recurrent problem with the spending has been the large amount of unspent revenue. This problem was partly addressed by broadening the scope of eligible spending and by improving project appraisal capacity. Unfortunately, there has not been a similar improvement in drafting annual spending programmes.

CONCLUSIONS

This product charge was the first economic instrument to be introduced after the political and economic changes of 1989. The difficulties encountered during preparation and negotiation were somewhat surprising for the environmental policy makers. Political will was a very important factor in jumping all the hurdles. Despite the time spent on preparation, the policy makers only recognized the need to carry out studies on the complex economic and business linkages of the transport fuel market after the first phase of implementation had already been completed.

There have also been problems on the spending side. The well-developed programme has been turned into an exercise of extreme earmarking. The annual spending programmes which became part of the CEPF annual spending guidelines targeted very narrowly defined areas, compared to the previously developed programmes. Combined with detailed spending rules and procedures, this resulted in slow disbursements and left large sums of money unspent at the end of several years. There was, for example, a programme component for retrofitting catalytic converters in two stroke engines, but the CEPF rules could not accommodate the administration of subsidies for very many vehicle owners, so the money allocated to this component was not spent for a long time. While these problems were the consequences of the actual practice of earmarking, and were not inherent in the idea of utilizing a spending programme approach to CEPF disbursements, they have still resulted in discrediting the notion of spending programmes.

The transport fuel product charge has proved to be an efficient revenue-raising mechanism. Revenue flow is predictable and is increasing steadily. There is no social welfare loss, in an economic sense, associated with its implementation. Administration costs are kept low by using the existing tax and custom duty collection mechanisms.

NOTES

1   Budapest University of Economics, Department of Microeconomics.
2   National Environmental Protection Programme, KTM, 1996.
3   These ambient concentration standards coincide substantially with the WHO standards.

4   See: Raucher et al (1992); HUF=Hungarian Forint; in 1996 approximately HUF153 = US$1.

5   The other products bearing an environmental charge are: packaging material, batteries, refrigerators/refrigerants, and tyres.

6   More details of the programme are presented in Annex 1.

7   The most well-known case of social pressure was the action by taxi drivers, who blocked and paralysed Budapest's traffic for three days, demanding lower gasoline prices. The three-day long negotiations between the democratically-elected government and representatives of the taxi drivers were televised. It was interesting to watch how the negotiations turned towards the issue of liberalizing the oil products market. While the government yielded to many of the demands with regard to liberalization, in the end liberalization actually resulted in substantial *increases* in gasoline prices.

8   Calculation of market share is based on data from the Central Statistical Office, 1995.

9   The analysis was done as part of the USAID technical assistance programme and was prepared by R S Raucher and C V Lula, 1992.

10 This summary is based on the supporting studies conducted for the product charge proposal, and on information in Raucher et al (1992). Note, in 1991 approximately HUF75 = US$1.

REFERENCES

Central Statistical Office (1995) *Statistical Yearbook of Hungary, 1994*, Budapest.

EGI (1997) EGI: Energy Sector Study. Report prepared for COWI Project Office, Budapest. Mimeo.

Government budget proposals, 1993,1994, 1995, 1996, 1997, 1998.

Ministry of Environment and Regional Policy (1996) *Guidelines for the Environmental Product Charges*, Budapest.

Ministry of Environment and Regional Policy (1994) *Air Pollution Indicators*, Budapest.

Ministry of Environment and Regional Policy (1996) Nemzeti Környezetvédelmi Program, (National Environmental Protection Programme), Budapest.

Ministry of Environment and Regional Policy (1993, 1994, 1995, 1996) Támogatási irányelvek, tervezetek (Spending Programme for the Environmental Product Charges), Budapest.

National Environmental Protection Programme, KTM (1994) *Air Pollution Indicators*. Ministry of Environment and Regional Policy.

Raucher, R S, Lula, C V and Trabaka, E J (1992) A Benefit–Cost Analysis of Mobile Source Emission Control Options under the Auto Fuel Charge Proposal for Hungary. Ministry of Environment and Regional Policy, Budapest, mimeo.

ANNEX
COMPONENTS OF THE THREE YEAR PROGRAMME ON TRANSPORT FUEL EMISSIONS REDUCTION[10]

- Two stroke engine programme:
  - installing new engines in 10,000 vehicles, at a total cost of HUF700 million (US$9.2 million);
  - retrofitting catalytic converters in 140,000 vehicles, at HUF1,800 million (US$23.7 million);
  - retiring 15,000 vehicles, at HUF 500 million (US$6.6 million).
- Bus programme:
  - installing of 'smoke reducers' on 5,000 buses at a total cost of HUF300 million, equivalent to HUF60,000 (or US$790) per bus;
  - installing 'turbo filters' on 2,500 buses at a total cost of HUF250 million, equivalent to HUF100,000 (or US$13,150) per bus;
  - installing modern engines in 400 buses, at a total cost of HUF250 million, equivalent to HUF625,000 (or US$$8,217) per bus.
- Four stroke engine programme:
  - installing 45,000 catalytic converters in 4-cycle vehicles, at a total cost of HUF600 million (US$7.9 million), which is 5.7 per cent of the total costs of the overall three-year programme.
- Infrastructure – urban transport programme:
  - road construction and repaving of 170 km, at a total cost of HUF1,300 million (US$17.1 million), equivalent to HUF7.65 per km (or U$167,500 per mile);
  - parking lot construction, at a total cost of HUF400 million, to accommodate 8,000 cars;
  - traffic controls, at a total cost of HUF300 million, to install 5 or 6 systems to improve traffic flow patterns.
- Infrastructure – noise protection programme:
  - construction of 8 km of noise barriers, at a total cost of HUF160 million, equivalent to HUF20 million per km (or US$438,000 per mile).
  - acoustic improvements to buildings (eg schools) located along noisy traffic areas, at a cost of HUF190 million for 320 flats, equivalent to HUF594,000 (or US$7,800) per flat.
  - tree and vegetation planting along roads, at a cost of HUF250 million for 500 km, equivalent to HUF500,000 per km (or US$11,000 per mile).
- Other infrastructure programmes:
  - constructing bicycle paths, at a cost of HUF300 million for 180 km, equivalent to HUF1.7 million per km (or US$36,500 per mile);
  - traffic reductions in residential areas (in medium and larger settlements), at a cost of HUF300 million for 15 to 20 areas, equivalent to HUF17 million (or US$225,400) per area.

- Public transport programme:
  - acquisition of 50 new vehicles, at a cost of HUF400 million, equivalent to HUF8 million (or US$105,200) per vehicle;
  - subsidies for mass transit fares, targeted to key areas and/or population segments, at a total cost of HUF500 million. These are expected to cover 23 million trips, so the average tariff subsidy would amount to HUF21.7 per trip
- Alternative fuels programme:
  - a set of subsidies for natural gas and agricultural-based bio-fuel production, at a total cost of HUF700 million (US$9.2 million), which is 6.6 per cent of the total costs of the overall three-year programme.
- Waste management programme:
  - disposal and recycling of transport-related residuals, at a total cost of HUF400 million (US$5.3 million) to support efforts to reduce land disposal damages.

14

WATER POLLUTION CHARGE IN POLAND

Zsuzsa Lehoczki[1] and Jerzy Sleszynski[2]

BACKGROUND

Freshwater resources are scarce in Poland. In fact, the quantity available – 1500 cubic metres per capita – is one of the lowest in Europe. This is mostly due to a low level of precipitation. The limited amount of surface water is responsible for 80 per cent of the water supply.

Water management, as in most Eastern and Central European countries, has a long tradition in Poland. The first water pollution charge appeared in 1965, and while there have been many amendments and changes since, most of these have affected only the charge rate, and the basic characteristics of this original charge remain in place today. The instrument was introduced in its present form in 1976 and the legal foundation of the present charge system rests with the Legal Water Act of 1974, which regulates both water quality and quantity management. The freshwater resource scarcity has meant that the quantitative aspect has dominated water management and policy, and in the meantime serious water pollution problems have developed.

In 1992, 60 per cent of Poland's total river length was classified as over-polluted according to physical-chemical parameters, and 88 per cent according to biological parameters. The most significant pollution problem is the very high level of biological oxygen demand (BOD), which is the result of a large volume of untreated sewage. High concentrations of heavy metals are another problem and there are also serious water quality problems along the Polish coastline (OECD, 1995). Untreated wastewater discharges are the major source of surface water pollution problems (Table 14.1 shows the amounts of these discharges over recent years). In 1989, about one-third of the wastewater was discharged without any treatment into the surface water, and only about 32 per cent of the discharges were treated at the required level (Nowicki, 1993). Where the effluent had been treated, the treatment efficiency was rather low, with only 40 per cent of the treated waste achieving 75 per cent BOD removal (OECD, 1995).

Table 14.1 *Direct Wastewater Discharges into Surface Water (billion cubic metres)*

Discharges from:	1993	1994	1995	1996
Industry*	7.7	7.8	8.1	8.3
Municipalities	2.0	2.0	1.9	1.8
Total	9.7	9.8	10.0	10.1

Note: * excluding cooling water and discharges into the municipal sewerage networks
Source: OECD, 1995

Industrial discharges are largely to blame for toxic pollution. Coal mining creates a particular problem of saline water in Upper Silesia, in the Vistula and Oder rivers. The coal mines in these areas discharge large quantities of salt, causing high concentrations of chlorides in these two rivers. The Vistula river is a major contributor to the pollution load of the Baltic Sea (OECD, 1995).

The National Ecological Policy document of 1990 specified medium-term priorities for the protection and rational use of water resources. This included a 50 per cent reduction in the river pollution coming from industries and municipalities, by the year 2000. Poland is a party to the Baltic Sea Action Programme and high priority was assigned to water pollution abatement efforts in the coastal zone. The revised National Environmental Policy Implementation Programme now specifies further reduction goals, including a 30 per cent increase in the volume of waste-water which is treated biologically and chemically. This, it is hoped, will reduce the Baltic Sea pollution load, from direct discharges and from rivers, by 20 per cent for BOD_5 and by 8 per cent for phosphorus. The set of medium-term priorities in the National Environmental Policy Programme will require the construction of an estimated 1,700 munici-pal, industrial and farm wastewater treatment plants, with a total capacity of 4.5 million cubic metres per day, and the reduction of salt water load discharged to the Oder and Vistula Rivers by 870 tons per year. These undertakings will need PLZ 68,500,000 million (US$3014 million) during the period of 1994–2000.[3]

These fairly ambitious goals were accompanied by changes in regula-tions, including an increase in the water pollution charge rate. In 1990, the unit charge rate for most polluting components was increased fifteen-fold.[4] These drastically increased unit charge rates were intended to be a major tool in achieving the policy goals.

WATER POLLUTION CHARGE

The water pollution charge is designed as a classic pollution charge. The payment is based on the total volume of selected pollutants present in the wastewater being discharged into surface water. The charge is levied on six major classes of pollutants (biochemical oxygen demand, chemical oxygen demand, suspended solids, heavy metals, chlorate and sulphate ions, and volatile phenols in wastewater) and with respect to critical limits set on wastewater temperature, radiation, and pH.

The water pollution charge is a price to be paid on every unit of effluent released into the water. Originally, despite low initial rates, the assumed environmental objective was a general incentive against excessive water pollution.

The water pollution charge is closely connected to a system of facility permits. Two separate permits are required: one for water intake and one for wastewater discharge. These two permits are issued together, and the permit for wastewater discharge requires an environmental impact statement, in the same way as the air pollution permit does. According to the Legal Water Act, wastewater discharge permits must be obtained by all enterprises that discharge their sewage directly into surface waters or soils, and all industrial and municipal sewage treatment plants.

The applicant must submit a detailed environmental impact statement approved by an independent reviewer. The statement must specify the different types of water effluent, their quantities and pollutant loads. Since provincial authorities are responsible for meeting the national ambient water quality standards (which are specified by a ministerial decree), it is their environmental protection departments which set the allowable pollution discharge levels for each source. The permit specifies the allowable amount of sewage that can be disposed, the highest allowable concentration of particular pollutants and other technical features of the sewage (such as limits on radioactivity levels or temperature).

As was seen in the air pollution permit system, the provincial authority takes into consideration the contribution of the applicant's wastewater to the surface water quality, using the national surface water quality standards as a general reference point. The provincial environmental protection department liaises with the Provincial Sanitary Inspectorate before issuing the permit.

Every discharger of wastewater who is required to have a water permit is also subject to water pollution charges. Polluters discharging directly into the sewerage system pay a user charge proportional to the operation costs incurred by the owner of the sewerage system. Apparently, the sewerage system owner/operators bear all further responsibilities for wastewater treatment and final discharge into the surface waters.

Non-compliance fines are levied on each ton of pollutant load exceeding the permitted level. The fine is calculated on the basis of a rather complicated formula taking into account the quantity of pollutant overload and the scale of violation for the following standards for the wastewater: the allowable temperature of discharged cooling water; the allowable acidity limit for the effluent; and the allowable level of radioactivity of the effluent. For social reasons, certain users, including some municipalities, pay much less for some categories of pollution – some times as little as one-tenth of the regular charge.

IMPLEMENTATION OF THE CHARGE

The charge payment is calculated on the basis of self-reported emissions. General exemptions are made for minor polluters (ie those whose pollution charge would be under US$93 (PLZ 2,370,000) and non-economic users of water resources. Provincial inspectorates are responsible for monitoring compliance with facility permits and verifying the accuracy of the pollution levels reported by the dischargers.

Table 14.2 shows the evolution of effluent charge rates over recent years. The most dramatic increases occurred in July 1990 and January 1991. The rates presented in the table are regular charges. However, the charge rates are differentiated in two ways: (i) by geographical regions (for instance, Katowice region experiences rates which are 100 per cent higher than in the rest of Poland);[5] and (ii) by type of industry. Regular rates are doubled for industries in polluter group no 1: chemical, fuel processing, metallurgical, machine, and light industries. The multiplying factor for enterprises in polluter group no 2 – ie the paper and pulp industry – is 0.85 of the regular rate. Food processing industries, which make up polluter group no 3, are charged half the regular charge rate. For polluter group no 4 – urban and rural municipal sewage, plus hospitals and social care institutions – the multiplying factor is 0.2 of the regular charge rate. The regular charge rate applies to all other polluters, which make up group no 5. Table 143.3 illustrates how the unit charge rate varied across these polluter groups, for BOD_5 load, in two different years.

Table 14.2 *Unit Charge Rate for Pollutants in the Water Pollution Charge (US$/ton)*

Year	BOD_5	COD	Suspended solids	Heavy metals (total mass)	Chlorate and sulphate ions
January 1990	11.2	6.8	1.0		0.5
July 1990	167.3	101.0	15.8		7.9
1991	524.4	299.8	46.8		23.6
1992	712.7	407.3	63.9	7338.2	40.4
1993	535.4	306.0	48.0	5513.9	30.4
1994	704.0	396.0	61.6	7040.0	39.6
1995	816.7	463.0	70.9	8250.6	46.3
1996	991.8				

Source: Sleszynski (1996)

Table 14.3 *Unit Charge Rate for BOD$_5$ in the Different Polluter Groups (US\$/ton of BOD$_5$)*

Polluter Group	1994	1996
Polluter group no.1	1408.0	1983.5
Polluter group no.2	598.4	843.0
Polluter group no.3	352.0	495.9
Polluter group no.4	140.8	198.4
Polluter group no.5	704.0	991.8

Source: Sleszynski (1996)

An annual adjustment procedure is a regular element of the implementation of the charge, in order to reevaluate all rates according to the anticipated inflation rate. Slippage in charge rates due to inflation is largely avoided by adjusting nominal charge rates to account for the predicted level of inflation in the next year. If actual inflation deviates significantly in either direction from the predicted level, adjustments can be made in the following year.

Environmental charges are collected once a year by the provincial administration environmental protection departments. As a result of a 1990 amendment to the general legal act, the Governor of the province may request large enterprises to pay their charges in quarterly instalments.

Enterprises are able to treat environmental charges as normal business expenses and to deduct the amount of charges paid from their taxable income. Thus, charges are treated as a normal production cost. An interesting provision allows enterprises to deduct the amount of the charges *levied* in the current year from the current year taxable income even if they are delinquent in making payments and don't actually *pay* the charges until the next calendar year.

Annual revenue collected from water pollution charges is somewhat moderate in comparison to other revenue sources. The revenue varies greatly from year to year, because of changes in charge rates and changes in the effectiveness of collection. The total water pollution charge revenue collected in recent years is shown in Table 14.4.

Table 14.4 *Annual Revenue from Water Pollution Charge*

Year	Charge revenue in million PLZ	Charge revenue in million US\$
1990	63,771	6.71
1991	1,132,874	107.05
1992	1,423,445	104.43
1993	1,357,630	74.82
1994	2,074,793	91.29
1995	2,540,000	104.96
1996	2,870,000	106.30

Source: Sleszynski (1996).

The revenue collected is specifically earmarked for expenditure on water pollution control and is divided among three different environmental funds in the following way:

National Fund for Environmental Protection and
Water Management (NF)   36%
Regional Environmental Funds of the affected region[6]   54%
Municipal Environmental Fund of the affected settlement[7]   10%

The revenue raised supports the operation of different subsidy schemes offered by the environmental funds.

In theory, water pollution discharge without a valid permit is penalized according to the quantity and impact of the pollution, though in reality fines are rarely imposed. Minor violations (ie those not exceeding a US$118 minimum fine) are excused, and many other polluters still operate without an emission permit because provincial authorities have limited resources to process or revise the permit applications. In practice, polluters of this category pay regular rates instead of the penalty rates designed for dischargers not holding valid permits.

DESIGN AND IMPLEMENTATION ISSUES

At present, the water pollution charge seems to be purely a revenue-raising tool, even though this is not made explicit in the official documents. The unit charge levied was originally based on general assumptions concerning different pollutants' harmfulness and environmental impacts, without any relation to the cost of the damage incurred. In principle, only *relative* differences in rates are correlated with differences in pollutants' toxicity or their potential to cause environmental damage. For instance, discharging heavy metals is assigned an extremely high rate. The *absolute* magnitudes of the rates are not set to reflect marginal damages to health and the environment or to correspond with the marginal costs of abatement. In practice, charge levels are set based on the intention that the amount of revenue collected should correspond to the investment needs of water resources management facilities, and primarily to the needs of wastewater treatment plants. However, this revenue-raising aspect has not been officially acknowledged. Therefore, the investment needs of water resource management have not been compared to the revenue raised, and so the unit charge rate has not been adjusted to fit the required financial resources.

The Polish environmental charge system is basically the only one in the region that has tackled the inflation problem successfully. The automatic annual revision of the charge rate is a legal requirement. Back in 1989, environmentalists managed to introduce legislation pegging the rates to the official inflation index. Unfortunately, the proponents of this measure did not realize that charges were pegged to the *ex post* inflation index, which only becomes available by the middle of the year after the one in which the charges have been applied. Therefore, in 1992 they switched to the present system, where the *expected* inflation rate is used for adjustment.

Earmarking water pollution charges for the water sector is the usual practice in most countries utilizing such instruments. In Poland, wastewater charges and fines are dedicated to financing efforts to reduce wastewater discharges, but not necessarily to address specific types of wastewater problems. Only revenues from mining wastewaters (ie saline water) are earmarked in this way, and must be used for the exclusive purposes of reducing saline mining discharge waters.

Water Basin Agencies (WBAs) were established in February 1991, to supervise water management in river basins, but unfortunately the present Legal Water Act does not allow these WBAs to collect or invest the water pollution charge revenue. Thus, WBAs are purely budgetary bodies, deprived of financial rights and responsibilities in water resources management. So far, WBAs' responsibilities are limited to providing balances of water resources, undertaking impact assessments for new investments, and conducting research studies on water withdrawal and wastewater discharge in their water basin. The new water act currently under discussion in parliament proposes changes that would allow WBAs to collect and retain at least part of water pollution charges. Such changes are opposed by many environmental policy makers, who want to keep this revenue source within the environmental funds system.

According to a 1993 report by the State Inspectorate for Environmental Protection, the following environmental impacts have been achieved by listed polluting enterprises (relative to 1989 levels): (i) the quantity of sewage discharged decreased by about 37 per cent; and (ii) several enterprises from the 'list of 80' most polluting enterprises were removed from the list and only a few new ones were added. At least a part of these improvements can be attributed to a decrease in the economic output of these enterprises in the years 1990–1993. However, most of the improvements can be linked to actions undertaken by enterprises involving technological changes, process modernization, and the installation of pollution abatement equipment. These actions have been responses to the entire set of water pollution control policy measures, including the pollution charge.

Despite these obvious achievements of the water pollution control policy, there are strong signs of enforcement and compliance problems. There are three ways in which enterprises fail to comply with the water charge regulations: firstly by violating the rules related to acquiring a water permit; secondly by falsifying the pollution data in their self-reports; and thirdly by not paying the charges that are levied.

Most of the larger facilities in Poland must operate with either a valid or temporary permit. In 1992, 17,389 facilities were registered as water polluters. Although they are all required to apply for facility permits, it is estimated that nearly half of these facilities operate without one. The backlog is largely attributable to limited local resources for processing permit applications.

The ability of the provincial inspectorates to verify self-reported discharges and impose fines is limited by staff resources and compounded by the large number of facilities and individual pollutants that must be checked. And a number of small and medium-sized enterprises do not seem to be included in the charge system at all.

The level of compliance seems to vary enormously from year to year. In particular, collection of the BOD_5 charge is an example of declining effectiveness of revenue-raising policy. Table 14.5 shows the percentage ration of the sum of charges already paid over the total amount of charges imposed. This ratio has fluctuated around 50 per cent since 1992. Note that it may be higher than 100 per cent where delayed payments from the previous years add to the regular payments.

Table 14.5 *Compliance with Charge Payment Obligations (BOD_5)*

	1990	1991	1992	1993	1994
Compliance ratio (%)	108	69	50	53	45

Many dischargers are experiencing economic problems which have severely limited their ability to fulfill their financial obligations. This is particularly apparent in the case of the coal mining sector, with many mines refusing to pay the charges and fines.

Table 14.6 shows a downward trend in the total amount of wastewater discharged. It is impossible to distinguish the impact of the water pollution charge on water quality from the effects of other policy measures and economic changes.

Table 14.6 *Total Wastewater Discharges in Poland*

	1985	1990	1992	1993	1994
Waste water (million m³)	12,903	11,386	10,048	9738	9797

Table 14.7 shows a considerable improvement in the wastewater intensity, calculated as the amount of wastewater in cubic metres per million PLZ of GDP.

An evaluation study of the environmental charge system in Poland looked at the distributional impact of the charges.[8] Unfortunately, the analysis did not distinguish between the different charges, but the overall assessment is that the private sector pays a higher share of the charges and receives a lower share in subsidies, in comparison to the public sector.

Table 14.7 *Wastewater Intensity*

	1988	1989	1990	1991	1992	1993	1994	1995	1996
m³/million PLZ	19.07	18.35	20.29	20.31	18.79	17.54	16.77	15.97	15.05

Source: Sleszynski (1996)

CONCLUSIONS

The role of the water pollution charge is to raise revenue. The marginal damages or marginal abatement costs of surface water pollution have not been considered when setting the unit charge rate. Recently, a regular inspection of environmental protection policy, provided by the Chief Board of Supervision, initiated a discussion on environmental damage as a possible basis for designing new rates for wastewater discharge. However, there seems to be a lack of research experience and available valuation methods, monitoring capacity, and political will to undertake such a revolutionary modification.

The unit charge rate for water pollution has been differentiated across different groups of pollutants in an attempt to make it more equitable. However, this solution impairs the efficiency of the charge and similar distortions can be observed in the surface and groundwater *withdrawal* charge system, where there is an even greater range of charge rates. These differentiations may need to be revisited, particularly in the light of the massive non-compliance rates for payment of these charges.

Researchers have suggested an increase in the charges on wastewater. On the one hand, they argue, the present rates are approximately six times lower than the level that would result in an efficient level of wastewater discharges.[9] On the other hand, the researchers are convinced that charges could be doubled without causing an unacceptable burden on households, who are implicitly paying a substantial part of these charges through user charge payments for wastewater services.

The major issue now is whether or not to authorize the WBAs to supervise complete water management, including collection of charges, and financing water treatment and water supply. This administrative decentralization would require the new Water Act to be submitted to Parliament for an open discussion. Environmental funds, however, will oppose any rechannelling of the water charges to WBAs unless the introduction of product charges, and in particular a fuel surcharge, will guarantee them a supplementary source of revenue.

NOTES

1   Budapest University of Economics, Department of Microeconomics.

2   Warsaw University, Economics Department, Warsaw Ecological Economics Centre.

3   PLZ = old Polish Zloty; PLZ 10,000 = PLN 1 (new Polish Zloty); in 1996, approximately PLZ 27,000 = US$1.

4   Unit charge on BOD increased from 11.1 US$/ton to 167.4 US$/ton; on COD from 6.8 US$/ton to 101 US$/ton; on suspended solids from 1 US$/ton to 15.8 US$/ton; on total mass of heavy metals to 7338.2 US$/ton and on chlorate and sulphate ions from 0.5 US$/ton to 7.9 US$/ton.

5   The term region here refers to what is sometimes called 'voidvodship', or 'wojewodztwo' in Polish.

6   There are 49 regional funds.

7   There are 2468 municipal funds.

8   The study was conducted by the Technical University of Bialystok.

9   The Cambridge economist A C Pigou introduced the notion of an economically efficient level of harmful effects (ie negative externalities). This is the level at which the social marginal cost is equal to the social marginal benefit associated with an additional reduction of the harmful effect.

REFERENCES

Directive of the Council of Ministers on charges on the use and alteration of natural environment, Dz U No 133, pos 638, December 27, 1993; amendments: Dz U No 51, pos 203, and Dz U No 140, pos 772 of 1994; Dz U No 153, pos 775, October 19, 1995.

Nowicki M (1993) *Environment in Poland – Issues and Solutions*, Kluwer Academic Publishers.

OECD (1995) *Environmental Performance Review of Poland*. Centre for Cooperation with Economies in Transition, OECD, Paris.

Sleszynski, J (1996) Two Case Studies on Implementation of Selected Economic Instruments in Poland. Paper submitted in the UNEP project on Economic Instruments, Warsaw.

Zylicz, T (1993) Case Study on Poland. Paper presented at the OECD Workshop on Taxation and the Environment in Economies in Transition, OECD, Paris.

15

WATER ABSTRACTION FEE IN HUNGARY

Ferenc Burger[1] and Zsuzsa Lehoczki[2]

BACKGROUND

Hungary lies in the centre of the water basin of the Carpathians – a mountain range running through the neighbouring countries to the north and east. A large proportion of Hungary's surface water supply therefore originates from outside the country. In fact only 145 m³/sec is generated inside the country while 2240 m³/sec comes from outside.[3] Hungary's groundwater resources are considered satisfactory, with an exploitable rate of deep groundwater resources estimated at 8.9 million m³/day (DHV, 1996).

Groundwater contributes a major part of the domestic water supply, while surface water extraction supplies other uses (DHV, 1996). The first water abstraction fee was introduced back in 1970, with the aim of adding non-direct budgetary sources to the finances available for water management. It was a fee for using water reserves (from either surface or groundwater supplies) and was channelled through an extra-budgetary fund called the Water Management Fund. The Fund financed the operation of 12 river basin water management directorates and provided subsidies for mostly municipal water supply, wastewater collection and treatment investments.

As market forces were becoming dominant during Hungary's transition to a market economy, several water use problems became apparent. One was the excessive use of karstic water (very good quality underground water) by the mining industry, which had resulted in a serious decrease in groundwater levels by the end of the 1980s. This problem has jeopardized the drinking water supply and the supplies available for medical needs and the tourist industry in several regions.

It also became apparent that the water user permit system was tying down an extremely large proportion of some surface water reserves (such as in part of the Kőrös region), while only a fraction of this water was actually being abstracted.[4] This introduced substantial uncertainties into water management and planning, and efforts began to try and reduce the gap between the volume of water reserved through water use permits and the real volume being utilized.

In 1992, a major overhauling of the water abstraction fee system took place and the effective fee rate is now differentiated according to the type of use. The charge rate has been increased for reserved but unused water abstraction, and there have also been several changes in the granting of exemptions.

WATER ABSTRACTION FEE

The water abstraction charge is an example of a natural resource rent.[5] It is a payment for the right to extract a natural resource (surface or groundwater), so it can be classified as an access charge. It is very close to an economic instrument called the natural resource use charge in the Hungarian framework law on environmental protection.

The basic legislation introducing the present system of water abstraction fee is a 1992 act on extra-budgetary funds. This act contains a chapter on the Water Management Fund and a water abstraction fee is defined as a revenue source to the Fund. The act sets the base rate for the fee and defines the policy on exemptions.

The aim, as is clear from the legal background, is to raise revenue for the water management tasks of the public administration. These tasks are defined by the Water Management Act of 1995 as: protecting against damage caused by water; ensuring the economic use of drinking water; encouraging efficient water use; and supporting the protection of water reserves. These tasks are to be carried out with the financial assistance of the Water Management Fund.

The regulations do not explicitly mention any incentive purpose for the instrument. But the legislation does state that multiplication factors must be combined with the base rate to calculate the effective charge rate. The multiplication factor should depend on the method of water use specification, the type of water utilization and water reserves, and the state of water supply management in the given region. The purpose of establishing the multiplication factors has been to encourage a water use structure that corresponds to the priority list of water usage given in the Water Management Act.

The implementation of the water abstraction charge is connected to the water abstraction permit system. All those who invest in and/or undertake the abstraction of water, whether from surface or underground supplies, must acquire a water permit. The permitted, ie 'reserved', quantity and type of water is specified in the permit and the permit applicant must also indicate the intended use of the abstracted water.

Since the main purpose of the instrument is revenue raising, its effect is connected to how the revenue is spent. Financing water management activities from the charge revenue is in line with the User Pays Principle. Economic efficiency can be increased if public water management tasks are financed by a related natural resource rent rather than by general taxation. Assessing the operation of the water abstraction fee, the effectiveness and efficiency of financing and subsidies through the Water Management Fund should also be looked at.

IMPLEMENTATION OF THE FEE SYSTEM

The fee must be paid by 'water users' and 'plant level water users'.[6] Water users are defined as those who carry out activities for which a water abstraction permit is required (regardless of whether or not they actually have the permit). Plant level water users are industrial units which abstract more than 10,000 m³/year from public waterworks which supply drinking water.

Exemption is given to those water users who use:

- surface water for ecological purposes (ie for protected areas);
- water for certain flood protection and other protection measures;
- water for irrigation during periods when a national drought has been declared;
- recycled water; and
- less than the threshold value (500 m³/year or 500 HUF/year).[7]

Exemptions are given to plant level water users if public health regulations require drinking water quality for more than 50 per cent of their water use (as is the case for some food processing plants).

The total payment is determined according to the following formula.

Payment = Water quantity * effective fee rate,

where: effective fee rate = multiplier * base rate

Water quantity is the actual volume of water abstracted if it is at least 80 per cent of the volume reserved in the water permit. If the actual volume abstracted is less than 80 per cent of the reserved volume then payment must also be made for part of the water reserved in the permit but not abstracted. In such cases, the water quantity in the above formula is 80 per cent of the reserved volume and not the actually abstracted water.

Effective fee rate depends on two variables: the base rate and the multiplying factor. The base rate varies according to compliance with the permitting process. It is higher if the abstracted quantity is above the permitted volume or if the water users are abstracting water without a permit. Table 15.1 shows the base rates which have been in effect since 1996.

Table 15.1 *Water Quantity and Base Rates for Calculating Abstraction Fee Payment*

Type of User	Level of Abstraction	Water Quantity	Base Rate			
			HUF/m³		US$/m³	
			1996	1998	1996	1998
Water Users	Abstracting less than 80% of the permitted (reserved) volume	80% of the permitted (reserved) volume	1	1.35	0.008	0.007
	abstracting between 80–110% of the permitted (reserved) volume	abstracted volume	1	1.35	0.008	0.007
	abstracting more than 110% of the permitted (reserved) volume	permitted volume difference between permitted and actual abstraction	1 2	1.35 2.7	0.008 0.016	0.007 0.013
	abstracting without permission	volume of abstracted water	5	6.8	0.04	0.03
Plant Level Water Users		volume of water bought from public waterworks	3	4.1	0.024	0.02

The value of the multiplication factor varies according to (i) the type of use for which the water is abstracted; and (ii) the total demand on the particular water deposit. There are different multiplication factors for surface and underground water abstraction.

In the case of *surface water*, the lowest multiplication value (0.001) is applied to in situ water use by hydro-power plants. The other end of the scale is 2.3, the multiplication factor for water use by economic activities in all but the energy sector. User categories between these two extreme values – in increasing order of the multiplier value – are:

- fisheries and rice plantations;
- irrigation (if there is no national drought);
- economic activities in the energy sector;
- public uses (household drinking and domestic uses, public institutions' drinking and communal water uses, and bath water uses).

There is much less variation in the value of the multiplication factor according to the total demand on the water deposit. There are four categories of total demand and for any given water use the difference between the two extreme categories of total demand is less than three-fold.

In the case of *ground water*, the lowest multiplication value is 0.5, for the lowest quality underground water used for public consumption or medical purposes. The highest multiplication value is 10, for the highest quality groundwater used for economic activities (ie for business purposes). Six different types of groundwater deposits are defined, with three quality categories for four of these types.

The usage categories (in increasing order of the multiplier value) are:

- medical purposes;
- public purposes;
- drinking water quality for business purposes (where public health regulations require drinking water quality for more than 50 per cent of their water use);
- other uses for business purposes.

The variation in the multiplier value is about the same for both water use and total level of demand, the ratio between the lowest and the highest values being 1:10.

It is clear from the above description that estimating even a range for an effective fee rate is not a trivial exercise. A very rough estimate is that the effective unit fee for permitted uses varies between 0.001 HUF/m^3 and 10 HUF/m^3.

Water users, after receiving the water abstraction permit, and plant level users, after starting to use water from public waterworks, must report to their Regional Water Authority. They must use a special form to declare their water abstraction or purchase of water. Each report must contain the basic data needed to calculate the effective fee rate and the total payment.

The water abstraction fee must be paid quarterly. Payment must be accompanied by a statement which details the actual quantity of water used, as this information is needed in assessing the fee level and payment obligations. The Regional Water Authority has the right and responsibility to check the validity of the reports and the payments. The scope of their authority with regard to payment collection is similar to that of the tax collection authorities.

The revenue is channelled to the Water Management Fund, which is an extra-budgetary fund. In 1991 the wastewater fines were redirected into the Central Environmental Fund from the Water Management Fund and since then the water abstraction fee is the only major revenue source to the latter.

Since 1993, 65 per cent of the Fund's revenue has been allocated to investment subsidies but this portion can also be used to cover debt service and repayment costs. Twenty-five per cent must be used to cover state water management tasks (such as flood control) and 2 per cent can be used to cover administration costs.

IMPLEMENTATION ISSUES AND IMPACTS

The need for a major overhaul of the water abstraction charge was recognized as a necessary response to the changing role of market price mechanisms and the scarcity of water resources. Substantial work was done to assess the impact of the very preferential treatment given to mining and heavy industrial users, particularly at the regional level. Even though it was not explicitly stated, a fee increase for those users was one objective of the revisions. However, the preferential access charge for water use was not the only subsidy that mining and heavy industry benefited from, and elimination of the different subsidies became the moving force for industrial restructuring. Therefore, the impact of the increased water abstraction fee is interwoven with the effects of other, probably more decisive, forces of restructuring.

Aggregate water consumption has steadily declined since 1990, as Tables 15.2 and 15.3 show.

Table 15.2 *Quantity of Water Sold (million m³)*

	1980	1990	1992	1993
Drinking water	792.7	888.9	841.0	808.9
Water for industrial use	64.0	66.1	63.3	60.4
Water for agricultural use	530.0	819.0	542.7	514.4
Total	1386.7	1774.0	1447.0	1383.7

Source: KSH, 1995

Table 15.3 *Quantity of Household Water Consumption (million m³)*

	1990	1993	1994	1995	1996
Household water consumption	579	474	445	412	396

Source: KSH, 1996, 1997

Considering that in 1992–93 Hungary's GDP and industrial production was still on the decline, we can assume that the instrument alone did not have a significant impact on the reduction of aggregate water use. However, the new fee system did have more influence on the structure and composition of sectoral water use. The proportion of underground water with a lower effective fee rate (ie with a more abundant supply) had increased somewhat by 1994. In the meantime, the industrial use of underground water supplies decreased by more than 20 per cent and this decrease is fully explained by the reduction in water outtake by the mining industry. A lack of detailed statistical data prevents a proper analysis to separate the fee's impact from other sources of economic restructuring.

For those users who were overestimating the water reserves needed, making the level of their payment dependent on the reserved quantity of water as well as on the actual quantity of water outtake did provide an incentive for more realistic estimates of their water needs. This effect can be demonstrated by the ratio of actual water demand to reserved supply.[8] Prior to 1990, this ratio fluctuated between 50 and 60 per cent for surface waters. In 1991 it was 49 per cent, by 1992 it had increased to 61 per cent, peaking at 67 per cent in the subsequent year. First and foremost, it was the food industry that responded with a reduction in the reserved quantity, but there are examples of cuts of as much as 50 per cent in the reserved quantity in other sectors.

The annual water abstraction fee payment is less than 1 per cent of total production costs in most sectors.[9] The most affected sector in industry has been the power sector, but even there it is not significant enough to create a competitiveness problem. Another potentially sensitive sector is agriculture, so the present system of multipliers for converting the base rate into the effective charge rate favours agricultural users to a large extent. The estimated ratio of water abstraction fee payment to total cost in agriculture is around 0.05 per cent. Overall, it is the municipal water sector which is most affected, with a cost ratio estimated at 0.9 per cent. In that sector, however, the impact has been softened by disbursements of relatively large sums of money for local governments' public waterworks.

The main role of the instrument is to raise revenue. Revenue targets have been implicitly defined through the statement that the Water Management Fund's resources must be sufficient to cover planned expenditures. Since the water abstraction fee is the only substantial source of revenue to the Fund, the estimated annual spending need of HUF2.5–3 billion (US$25–30 million) in the first phase, rising to a subsequent HUF3–5 billion (US$28–45 million), can be viewed as a revenue target for the fee. In fact, the unit fee rates were calibrated to create a revenue stream of approximately that amount.

The establishment of multiplier values was the subject of extensive policy discussions. The set of different multipliers is complicated and there have been some wide fluctuations. For example, in the power sector the multiplication factor for hydroelectric power plants was set at 0.0002 in 1992. Within a year, a ministerial decree had raised the factor to 0.05, and in another five months it was revised to 0.001.

Table 15.4 shows the annual revenue flow from the charge. The year-to-year variation in the revenue collected partly reflects changes in the water charge system and also some enforcement problems. In 1990–91 the unit charge varied between 0.06 and 0.9 HUF/m^3. The big increase in 1992 was due to the major overhauling of the system and the establishment of a base rate of 0.4 HUF/m³. In 1993, the slight decline in nominal terms (a larger one in real terms), was due to the introduction of more changes, including an increase in the base rate to 0.5 HUF/m³, and changes in the multipliers, which offset the rate increase and resulted in a decrease in the amount of revenue collected. In 1994–1995 the base rate was raised to 0.8 HUF/m³ partly with the intention to correct for the lower revenue collected in the previous year.

Increases in the base rate are sufficient to maintain the real value of the base rate since they are in line with the inflation rate. In fact, these increases are above the changes in industrial price indexes. No regular inflationary adjustment has been established in the legislation. The current practice is to adjust the base rate as part of the discussions on the annual state budget. That adjustment process means that adjustments to the basic fee rate can be made to allow for inflation, and to respond to changes in the revenue target.

Table 15.4 *Annual Revenue from the Water Abstraction Fee*

Year	Charge for water abstraction		Water abstraction fee	
	million HUF	million US$	million HUF	million US$
1990	896	14.2		
1991	850	11.4		
1992 (estimated)	2095	26.5		
1993			1976	21.5
1994			2900	27.6
1995			3213	30.6
1996			3724	29.6
1997 (expected)			3960	21.2
1998 (planned)			5150	25.1

Source: KHVM (MTCWM) reports and yearly government budget proposals

The resources that can be used for administration, enforcement and monitoring are limited to 2 per cent of the revenue. Since this 2 per cent must cover the entire Fund administration costs, including a system of applications for support and the administration of disbursements, there is not enough left for monitoring and practically none for evaluation. In 1995, around US$0.5 million (HUF60–65 million) was available for administration purposes. Two-thirds of it (about HUF40 million) was used by the regional water directorates, distributed between them according to the number of water users in each. The rest was spent on central administrative costs at the National Directorate of Water Management. This amount is sufficient to cover the administration costs of financial transactions but is not enough to make assessments or establish regular mechanisms to evaluate the performance of the fee as a policy tool.

The regional water directorates, as fee collecting authorities, have a status similar to that of the tax authority and the rules for payment collection are even stricter than those in operation for tax payment. The law does not allow for the postponement of payments, deferred payments, or reductions in the payment obligation. While the enforcement of payments from users with valid permits is considered satisfactory, there are significant problems with the identification of unauthorized water users. Detecting water users who do not have valid permits would necessitate much more work in the field than can be financed from the revenue available for administration.

During inspection visits by the regional directorates, the water use category stated by some water users has occasionally been challenged and subsequently overruled by the National Directorate of Water Management. Since the reported water use category determines the multiplication factor and so can create large differences in the effective fee rate, the categorization has a large impact on the size of total payment obligation. In the new market and democratic setting, such overruling on the type of multipliers is appealed by the water users. The most important classification controversy has just appeared before the public administration court. Nevertheless, the need for a more accurate interpretation of the law has not yet emerged.

Since the water abstraction fee is a revenue-raising instrument, it is worth looking at how the revenue is being spent. Table 15.5 shows the amount of investment subsidies allocated to municipalities and industry through applications. The table shows that support for municipalities dominates the spending. Considering both the revenue-raising and disbursement side of the fee, we can see that municipalities are the net receivers and industry is the net payer.

The trend in spending shows an increase in the funding of the overall operation costs of the water directorates. The share of such funding, above the 2 per cent reserved for administration costs of the Water Management Fund, has increased from 20 per cent to 25 per cent. The share of subsidizing water-related investments has decreased from 70 per cent to 65 per cent since 1993.

Subsidies are allocated through an application process. A call for proposals is issued annually, developed jointly by the relevant ministries. Water protection is gradually being promoted as a priority investment area. The share of subsidies going to sewage collection and treatment increased, while the support for investments in water supply became negligible by 1996.

CONCLUSIONS

During preparation for the changes in the water abstraction fee system in 1992, policy makers emphasized that the previous system did not satisfy society's growing expectations for rational utilization of water supplies, nor did it reflect the increasing value of water as a scarce natural resource. Even though it has been stated in some policy documents that

Table 15.5 *Subsidies from the Water Management Fund (million HUF)*

Year	Approved for municipalities		Approved for industrial water users	
	Total	Out of which, subsidies for sewage and wastewater treatment	Total	Out of which, subsidies for water saving and quality protection
1992	507	177	97	71
1993	919	569	34	34
1994*	840	410	0	0
1995*	1086	902	40	40

Note: * These numbers do not include payments to projects approved in 1992 and due in 1994 or 1995.
Source: Government budget proposals, 1992, 1993, 1994, 1995

the water abstraction fee must be viewed as a type of rent, which reflects the intrinsic value of this scarce natural resource, such resource valuation has not happened. The multiplier values were set to allocate the overall revenue target, largely on the basis of the burden-bearing capacity of the sectors. Some consideration has been given to the relative scarcity of different water supply sources.

The multiplication factors occasionally resulted in fees high enough to motivate some industrial corporations to either switch from the use of relatively valuable subsurface water supplies to alternative sources, or to increase their water recycling. It is clear, however, that most of the environmentally beneficial trends in water use have been due to the overall economic restructuring and to water price increases (due to the elimination of subsidies), rather than to increases in water abstraction fees.

The water abstraction fee serves the intended revenue-raising objective fairly efficiently with reasonable administration costs. A full economic welfare assessment of the instrument would require valuation of the water resources. Such a valuation exercise could guide the calibration of the unit fee to reflect the *absolute* scarcity of water resources. Scarce water supply, on the other hand, is a problem in only some regions. During the drafting of the new water management law in 1993–94, the idea of tradable water extraction permits was raised, for certain types of water supply in certain regions. The idea was never developed into a detailed proposal, but there is still the possibility of using such a complementary instrument in the future to enhance the effect of the water abstraction fee.

NOTES

1   EXTERNAL Environmental Economics Consulting Ltd.
2   Budapest University of Economics, Department of Microeconomics.

3   2240 m³/sec and 145 m³/sec are values for the flow of 80 per cent probability in August, which is the base figure used for water management purposes.

4   There was no similar problem for underground water because the main users of that water supply were municipal waterworks, whose needs are easier to predict that those of industry.

5   As the term natural resource rent is used by T. Panayotou in his papers on economic instruments and financing mechanisms (Panayotou, 1994).

6   These plant level users correspond roughly to what used to be known as 'big industrial users'.

7   HUF= Hungarian Forint; in 1996, approximately HUF153 = US$1.

8   See Ministry of Transport, Telecommunication and Water Management, 1992.

9   A fairly simple economic impact assessment was attached to the proposed new legislation on the Water Management Fund.

REFERENCES

Central Statistical Office (KSH) (1995) Magyar Statisztikai Évkönyv 1994 (*Hungarian Statistical Yearbook 1994*).

Central Statistical Office (KSH) (1996) *Major Public Utilities Indicators*.

DHV (1996) Baseline Report for the Water Quality Management project. Mimeo.

Government budget proposals, 1989, 1990, 1991, 1992, 1994, 1995, 1996.

Ministry of Transport, Telecommunication and Water Management (1992) Proposal for the new act on Water Management Fund, Budapest.

Ministry of Environment and Regional Policy (1996) Nemzeti Környezetvédelmi Program (National Environmental Protection Program).

Panayotou, T (1994) Economic Instruments for Environmental Management and Sustainable Development. Report prepared for UNEP. Harvard Institute for International Development, Massachusetts.

Part IV

Latin America
Case Studies

16

WASTEWATER EFFLUENT CHARGE IN MEXICO

Ronaldo Seroa da Motta,[1] *Hugo Contreras*[2] *and Lilian Saade*[3]

BACKGROUND

Water pollution is a major health hazard in Mexico and is estimated to cost approximately $3600 million dollars.[4] The pollution load is concentrated in a few catchment areas where urbanization and industrialization levels are high. About 60 per cent of the country's wastewater discharge is generated in the Federal District and seven of the 31 states, where 55 per cent of the Mexican population is located.

At least one third of the 79 overexploited aquifers of the country show irreversible damage. Despite the importance of recovering treated wastewater to satisfy increasing water needs, the sewage treatment capacity in Mexico is very limited and currently less than 10 per cent of effluents are treated.

The Mexican government is using several economic instruments, in combination with control-oriented instruments, to effect changes in polluters' behaviour by internalising their environmental costs and benefits.

The wastewater charge, in place since October 1991 and recently revised, is an example of the use of economic instruments for water management in the country. The purpose of this charge is to enforce effluent standards and to provide an incentive to induce firms to invest in measures to improve the quality of their wastewater.

The original 1991 charge was a non-compliance charge, applied only when pollutant concentration levels exceeded standards. The 1995 revision went beyond that and turned the charge into a type of tax levied on all levels of concentration.

THE ORIGINAL CHARGE

The wastewater effluent charge is legally bound by the Mexican Constitution, the General Law for Ecological Equilibrium and Environmental Protection, The Federal Water Act and the Federal Water Rights Law. At a decentralized level, the municipalities are responsible for water supply and wastewater management. At the national level, the National Water Commission (the Spanish acronym of which is CNA) is responsible for the promotion and implementation of federal infrastructure and the necessary services for maintaining water quality. The charge is applied by CNA and the resultant revenue goes to the Treasury without returning to either the municipalities or polluters in the form of a fund.

In the 1991 version, the charge was levied on every cubic metre of water discharged and the charge level varied according to four zones, which were defined according to water availability criteria.

The charge was to be applied on all polluting sources that discharged effluents. For those sources exceeding a monthly discharge of 3000 m³, charge levels were determined by the product of the concentration level times the zone charge. Monthly discharges of less than 3000 m³ were charged a flat rate, the total fee relating only to volume levels.

Only two pollution indicators were considered: the concentration of chemical oxygen demand (COD) and total suspended solids (TSS).

The Federal Water Rights Law conceded charge exemptions in the following cases:

- those who complied with the effluent standards;
- those discharging wastewater into drainage or sewer systems that were not state-owned; and
- those users holding a certificate issued by the CNA indicating that the water was used and returned in the same condition.

The 1991 version of the effluent charge was clearly a non-compliance charge as it was applied only to concentrations over standard levels.

THE REVISED CHARGE

In December 1995, revisions to the Federal Water Rights Law were introduced, setting new procedures by which the charge payments were to be calculated. From then on, the charge has no longer been based on water availability zones but on the different assimilative capacity of receiving water bodies, defined according to the current use of the water and the treatment level required for the amount of pollutants found in the water body. Three categories of assimilative capacity were defined to classify the country's water bodies:

- Type A: water bodies with a lower treatment level;
- Type B: water bodies requiring secondary treatment;
- Type C: water bodies requiring a more sophisticated treatment level.

The effluent charge now also varies according to different levels of pollutant concentration, as described below. To sum up, the charge structures, at 1995 prices, before and after the new legislation are:

Prior to 1996:

Criteria:
- Water Availability Zones (Zone 1 corresponding to areas with highest water availability, Zone 4 to areas with lowest availability);
- Concentrations of COD and TSS.

For every 1000 m^3 of wastewater disposed:

in Zone 1: US$78.7
in Zone 2: US$19.64
in Zone 3: US$7.77
in Zone 4: US$3.91

For pollutants discharged:
- for every ton of COD in the discharge:
in Zone 1: US$51.16
in Zone 2: US$12.7
in Zone 3: US$5.08
in Zone 4: US$2.5

- for every ton of total suspended solids:
in Zone 1: US$90.5
in Zone 2: US$22.6
in Zone 3: US$9.0
in Zone 4: US$2.5

The total charge value would be the sum of all the above charge components.

After 1996:

Criteria:
- Assimilative capacity of the water body (Type A requiring least treatment, Type C requiring most);
- Concentrations of *either* COD or TSS, depending on which pollutant is more concentrated in the wastewater.

The Federal Water Rights Law now establishes four categories of pollutant concentration:

a) If the concentration of the pollutant is above 150 milligrams per litre (mg/l):
   in Type A: US$0.77
   in Type B: US$0.77
   in Type C: US$1.60
b) If the concentration of the pollutant is between 75 and 150 mg/l:
   in Type A: (US$0.01 times the concentration of the pollutant) minus US$0.69.
   in Type B: US$0.77
   in Type C: US$1.60
c) If the concentration of the pollutant is between 30 and 75 mg/l:
   in Type A: US$ 0.04
   in Type B: (US$0.02 times the concentration of the pollutant) minus US$0.37
   in Type C: US$1.60
d) If the concentration of the pollutant is 30 mg/l or less:
   in Type A: US$ 0.04
   in Type B: US$ 0.09
   in Type C: (US$0.07 times the concentration of the pollutant) minus US$0.47.

The new charge system is stricter than the previous version as there are no longer any exemptions for emissions below the standards set for COD and TSS. The charge is no longer a non-compliance charge but rather can be regarded as a kind of water pollution tax levied on all levels of pollutant concentration.

This revision is significant in both economic and environmental terms. Firstly, because a permanent incentive was created for users to reduce the volume and concentration of their pollution, and secondly because the water body's assimilative capacity is now taken into consideration.

However, other forms of charge exemption have been created. Polluters with a monthly discharge volume of less than 3000m^3 will still have the option of paying a fixed flat charge according to the receiving water body. And, public water suppliers to municipalities with less than 2500 inhabitants have also been granted an exemption, to reduce the fiscal burden on small municipalities.

As a result of the economic crisis in Mexico and the high level of defaulting on effluent charge payments, the government published Presidential Decrees in 1995 cancelling previous debts from the old charge system in cases where pollution abatement measures had been undertaken, and subsequent charge payments had been made. And large debts, which could threaten the financial survival of the companies concerned, were also cancelled.

Another important exemption was introduced for those polluters who had not met the deadline for the construction of their pollution abatement facilities, but who had demonstrated to the CNA that at least 80 per cent of the construction work was completed. These polluters were allowed to pay the lower charge level levied on polluters whose construction work was complete. Their own works, however, needed to be finished and put into operation before October 1996.

Agricultural run-off is not subject to water charges, due to its diffuse nature, which would make effective enforcement impossible.

The Federal Law of Charges is revised every year. Since 1997, this law allows the introduction of economic incentives for users of water receiving bodies, who have adopted processes for better water quality than that required by the standards. There have also been some changes in the classification of certain water bodies and additional pollution indicators are now being considered.

For practical reasons, to control the pollution level from municipal and industrial sectors, it was considered more effective to formulate two official standards published in 1997 (one of them still subject to public consultation). These standards establish the maximum permissible levels of pollutants for wastewater discharges to national waters and goods, and the maximum permissible levels of pollutants for wastewater discharges (except for households) to sewerage. This represents an enormous effort to simplify the system, since this new approach collapses the previous 43 official standards (mostly for specific industry sectors) into only these two.

The main purpose of this new approach is to provide incentives for polluters to adopt new practices, processes and technologies for the reduction of their polluting emissions. This change was driven, among other things, by the fact that the government perceived a large default among users, associated with the economic crisis that had struck the country. It was, therefore, considered that a gradual approach, or multi-stage approach, was the best option. It has been established that the deadlines for complying with the maximum permissible levels for discharges to national waters are, for the year, 2000, 2005 and 2010, according to the population size in the case of municipalities and the range of biochemical oxygen demand for non-municipal discharges.

The current legal framework establishes penalties for non-compliance. It is too early, however, to fully evaluate the effectiveness of these new standards. In fact, in the past months there has been a reduction in collection mainly explained by the extension given to polluters for the construction of their abatement facilities. One important feature is that these standards mean that all dischargers into a specific water body can be controlled using the same parameters. This way it is possible to enjoy economies of scale while encouraging polluters to treat their discharges jointly.

This instrument has also the advantage of being dynamic. The parameters considered are the same as in the Federal Law of Charges and the Special Conditions of Discharge, therefore allowing for the homogenization of policies. There is still a gap, however, in the development of a new regulatory framework for sludge originating from treatment plants and for the establishment of maximum permissible levels of pollutants for artificial aquifers' recharge and the conditions for water infiltration to soil and subsoil, which the Mexican government is working on. A new standard authorizing the use of treated wastewater for public services has recently been released for public consultation.

IMPLEMENTATION ISSUES

The main intention of the CNA in introducing these changes was to force polluters to comply with the pollution control regulations, rather than to increase the revenue obtained from the charge.

The Federal Water Rights Law establishes that the procedure for payment of the charge is based on self-reporting, ie polluters determine for themselves the amount to be paid. That may partly explain the low level of revenue raised so far, as the monitoring and control systems have not been efficient. This problem relates to the lack of financial resources and infrastructure within CNA to supervise firms regularly.

However, the CNA is trying to be more selective in an effort to increase revenue and to improve its institutional capacity. Their strategy now is to carry out inspections mostly of industrial discharges and users in cities of more than 10,000 inhabitants (about 700 cities in Mexico fall into this category), where the highest concentrations of discharges are found.

Random monitoring visits are already undertaken. The CNA organizes teams of on average three people to visit different firms. However, the size of the office responsible for monitoring is small and most of the staff have other duties in addition to the monitoring visits. Although the effectiveness of the new charge system can not be analysed yet, the enforcement capacity of CNA to apply the wastewater charge represented only 2 per cent of the CNA's potential total collection level in 1994.

When an analysis was done to assess the possibility of increasing the charges to better reflect the real cost of treating wastewater, it was determined that a substantial increase would cause serious problems for sugar mills and sanitation companies.

However, revenue levels have increased with the help of improved monitoring and enforcement, as shown in Table 16.1. The increase in revenue from 1992 to 1995 was the result of certain measures undertaken by the CNA, including expansion of the register of charge payers, publicity directed towards users, and improvements made in the monitoring activities.

The money collected from the wastewater charge is diverted to the Treasury and CNA then receives a share as a budgetary provision of its services.

Table 16.1 *Revenue Level from Wastewater Effluent Charge (US$ million)*

Year	Revenue level
1992	4.98
1993	7.23
1994	9.19
1995	10.22

In addition to the ineffective application of the charge system, the various exemptions, which were previously granted, also reduced the ability of the charge to raise revenue and promote efficient water use during that period. Investments and operations in water distribution have been heavily subsidized. Since industry pays a higher contribution than other water users, this creates cross subsidies for the financing of these types of municipal water services.

As a result of this situation, water supply revenues have made little financial contribution and so water resource management is increasingly dependent on the federal budget.

CONCLUSIONS AND RECOMMENDATIONS

The main purpose of economic instruments in environmental management is to induce change in the pattern of natural resource use by internalizing environmental costs and benefits.

Although the potential charge revenue expected from the Mexican wastewater charge would provide the means to expand investments in pollution abatement measures in municipalities and firms, this may not be achieved in practice if charges remain low and their enforcement remains so partial.

With new legislation in force, the charge system is now more flexible and comprehensive, since charges can be applied regardless of emission standard levels. However, without proper institutional capacity building to improve the monitoring and control of discharge activities, enforcement problems will not be overcome and the potential of the charge system will not be fulfilled.

NOTES

1   Instituto de Pesquisa Economica Aplicada, Brazil.

2   Subsecretaria de Planeacion, Mexico.
3   Subsecretaria de Planeacion, Mexico.
4   See Margulis (1992).

REFERENCES

Colección Porrúa (1994) Constitución Política de los Estados Unidos Mexicanos. 102ª edición, México.

Comisión Nacional de Agua (1990) Estudio sobre cuotas por derechos de descarga de aguas residuales, Informe final, Subdirección de Planeación y Finanzas, México.

Comisión Nacional de Água (1994) Estudio sobre el Precio del Agua en México, Resumen.

Comisión Nacional del Agua (1994) Informe 1989–1994, México.

Comisión Nacional del Agua (1995) Planeación Nacional Hidráulica. Experiencias, Resultado y Perspectivas.

Consejo Nacional de Investigación (1995) El Agua y la Ciudad de México.

Ley de Aguas Nacionales y Regiamento de Aguas Nacionales (1992).

Ley Federal de Derechos en Materia de Agua (1994).

Margulis, S (1992) Estimativas dos Custos Ambientais no México; in: May and Seroa da Motta (orgs) *Valorando a Natureza*, Ed. Campus, Rio de Janeiro.

Quiroz, J (ed) (1995) Análisis Econômico de la Contaminación de Aguas en América Latina. CINDE, Santiago.

Saade, L (1998) New Approach to Wastewater Quality Control in Mexico. Paper presented at the 8th Stockholm Water Symposium, 10–13 August.

Secretaria de Desarrollo Social – Instituto Nacional de Ecologia (1994) Informe de la Situación General en Materia de Equilibrio Ecológico y Protección al Ambiente 1993–1994. México.

17

WATER POLLUTION TAXES IN COLOMBIA

Ronaldo Seroa da Motta,[1] *Guillermo Rudas*[2] *and Juan Mauricio Ramírez*[3]

BACKGROUND

Colombia has abundant renewable water resources with more than 1000 river systems and 720,000 micro-watersheds. Rainfall levels exceed 2000 mm per year for over 88 per cent of the country, with a national average of 3000 mm per year. The total volume of rainfall is 3425 km^3 and the evaporation volume is 1313 km^3, giving an overall water availability of 2000 km^3 per year.

Despite these large quantities of both surface and atmospheric water, access to safe water and adequate sanitation is far from optimal. Although 87 per cent of the urban population has access to house connections (62 per cent to good quality water), in rural areas the access to house system connections is only 33.4 per cent (and only 9.1 per cent to good quality water).

The differences in access to safe water and sanitation by income groups are great, especially in rural areas where a non-poor household is four times more likely than poor households to have a house connection and ten times more likely to have sanitation services. In urban areas the access of the poor to sanitation services is less than 70 per cent, compared with almost 98 per cent access for non-poor households.

The level of contamination of rivers and water bodies, which receive discharge from the principal industrial belts of the country, is particularly high. The contamination of the Bogota River starts very close to its source, with discharges from the tanning industry. Further downstream, a stretch of 150 kilometres receives sewage from several municipalities and waste from slaughterhouses, horticultural enterprises, coal yards, foundries, thermoelectric stations and other industries. By the time it reaches Bogota, the river is already carrying 35 per cent of its contamination. The city sewerage system runs into the river with no purification, through three tributaries: River Juan Amarillo, which discharges a daily load of 135 tons of suspended solids and 105 tons of organic matter; the River Fucha, which contributes 650 tons of suspended solids and 300 tons of organic matter every day; and the River Tunjuelo, which contributes a daily 571 tons of suspended solids and 80 tons of organic matter. Both the River Fucha and the Tunjuelo carry heavy metals and toxic substances such as cadmium, chrome, mercury, lead, detergents, fats and oils. The galvanizing and anodizing industries which operate in small workshops and garages throughout the city require large quantities of rinsing water to eliminate the chemical products used in their processes. The health costs associated with this water pollution are estimated at around US$17 million per year.

Bogota City Hall recently drew up a 20-year clean-up plan for the Bogota River. The project consists of installing a purification plant at the confluence of each of the three principal tributaries.

In Colombia economic instruments have long been applied for water management. Water use charges, for instance, have been in force since 1942. In 1984 a charge on water effluent discharge was introduced. However, both these experiences have failed to curb the degradation of water resources or to collect the necessary revenue to implement water management policies. In 1993 a comprehensive new pollution tax was imposed, incorporating water use and water pollution charges, in an attempt to make a real impact on water users' behaviour.

LEGAL CONTEXT

Water Use Charges

Regulation on water use charges in Colombia dates back to 1942 when the so-called 'vigilance' service rates were established by the Ministry of Agriculture to finance the monitoring of water use. From 1974 onwards, water use charges were introduced by Inderena (the Institute of Renewable Natural Resources) and the regional Environmental Protection Agencies (EPAs), who were made responsible for setting the charge level.

In 1982 Inderena established eight types of charge rate according to water usage, from hydroelectric generation (the cheapest one, at US$0.012 l/sec/month) to the use of water bodies for shipping (charged at US$20 per vehicle/semester).

These charges were set at low levels and had no significant impact on the cost of water to users (see Rodriguez-Becerra and Uribe, 1995) and were not systematically collected. There were, however, some notable exemptions, such as a regional EPA in Valle del Cauca, which in 1978 started to apply charges according to different water uses and achieved considerable revenue collection. Another successful case concerned the Tota Lagoon, where rate values tripled in 1990 and increased again by 70 per cent in 1993, resulting in a 50 per cent decrease in water consumption from the lake.

At the present time, in compliance with the principles set out in Law 99 (discussed in detail below), the water use charge is calculated on the basis of the *annual costs* of water supply investments and administration in water basins used for water supply, divided by the *total volume* taken from the water basins. Charge levels for users are calculated as an exponential function of the user's consumption.

Pollution Charges

Decree number 1594 of 1984 introduced a pollution charge which was a retributive charge to be collected according to the level of pollution discharge. These pollution charges however were simply revenue-raising mechanisms to finance the environmental authorities and have not been used as economic instruments to induce behavioural change among polluters.

Moreover, the application of these taxes has been very limited due to:

1   the judicial disputes which arose when the environmental authorities tried to enforce the charges, as users argued that these authorities, who would be the recipients of the charge revenue, were not providing adequate services; and
2   the exclusion of public sector companies from the charge system, including those public bodies responsible for waste water treatment.

The application of these charges has indeed been very limited. According to Rodriguez-Bezerra and Uribe (1995), available data for 1989 indicated that only US$116,000 was collected from a potential revenue of almost US$90 million.

Water Taxation in Law 99/1993

Law 99/1993 emphasized the use of economic instruments and in particular the use of compensatory and retributive pollution charges aimed at 'inducing natural resource users to comply with environmental standards, to modify their consumption patterns and to encourage the adoption of processes and the consumption of goods produced through clean technologies'.

Those changes enabled water pollution charges to be regarded as a kind of pigouvian tax rather than as a financing mechanism.[4] The compensatory tax and retributive tax were originally intended to be combined into one overall water pollution charge. The compensatory tax relates to the compensation required to ensure a given level of quality for a given water resource. Compensatory taxes are aimed at covering the expenses borne by environmental authorities to guarantee the renewability of resources. The retributive tax relates to the retribution or recompense to the water body for its acting as a deposit for waste.

The major modifications to the definition of retributive rates in the new legislation are:

1   the new tax must reflect the social costs of pollution and not simply the administrative costs of resource maintenance and control. These costs cover the negative impacts on both the community's welfare and on the natural resources;
2   taxes are applied for using the resource as a receiver of the polluting discharge, regardless of the function undertaken by the environmental authority; and
3   all activities, whether public or private, are subject to these taxes.

Law 99 does not allow the application of retributive pollution taxes on discharges exceeding standards set by the environmental regulation. Any discharge above these standards is subject to sanctions stated in the environmental legislation, such as fines and compulsory shutdowns. Therefore, retributive taxes will act as a complementary instrument to encourage polluters to reduce their discharges to below standard levels. Of course, the economic effectiveness of the tax depends on the setting of flexible discharge standards.

Two additional procedures are being undertaken which could widen the scope of this retributive tax as an instrument to encourage water treatment at source. Firstly, the Ministry of the Environment is currently promoting agreements with the productive sector to establish specific plans for the 'establishment of goals at the short, medium and long terms to [achieve the] compliance of the environmental parameters set'. These agreements allow enterprises to postpone their full compliance with the strict water quality norms, as long as they keep within the mutually-agreed schedule. In this context, the retributive tax can operate as an economic incentive for these companies to improve their compliance with discharge standards.

Secondly, there is the possibility of combining the retributive rates with other economic instruments such as non-compliance fines on those polluters who do not subscribe to agreements and who remain outside the parameters set by environmental standards. If these fines are established proportional to the magnitude of the pollutant discharge, they will perform an identical function to that of retributive rates.

The Ministry of the Environment defines the tax level and collection system on a yearly basis, by formulating a tax structure and listing the variables to be taken into account in its determination. The Ministry also establishes a minimum tax level applicable nation-wide. The local environmental authorities – rural and urban EPAs – are then responsible for adjusting the rate levels to reflect the particular conditions of each location. The local tax rate levels must be at least as high as the national level.

Recent Progress in Implementing the Retributive Tax

The actual implementation of the retributive tax finally got underway in 1997 following decree 901, which defined a minimum tax rate and the mechanism for adjusting the rate levels to reflect the particular conditions of each location. The main issues defined by this decree are the following:

- Initially the tax will be charged only on emissions of BOD and TSS. The minimum tax rate will be US$0.03 and US$0.013 per kilo respectively.
- Each regional EPA will establish an environmental target that will be revised every five years. The target will be determined within a consensus-based regional process involving the different actors and communities associated with the water resource in question.
- Polluters must present an emission report every six months. EPAs will organize random emission tests to verify the accuracy of the information.
- Tax rates will be adjusted every six months on a regional basis. The tax rate can only be adjusted until the predetermined regional goal is achieved.
- Tax levels will be adjusted by increasing the regional factor (a multiplier) by 0.5. This means that for those regions adopting the minimum level set by the Ministry, the tax will double the first year, and increase by 50 per cent, 33 per cent, 25 per cent and 20 per cent between the second and fifth years.

As for the compensatory tax, plans to implement it have been put on hold, partly because of the strong political opposition that is expected, given the recent implementation of the retributive tax.

IMPLEMENTATION ISSUES

The following sections will concentrate solely on the retributive aspect of the pollution tax since this is the most ambitious aspect of the instrument and, consequently, presents very interesting problems at the implementation level. It should be noted that this tax has just recently begun to be enforced, so these problems have not yet been fully solved.

Coverage of Polluting Sources

To be able to operate as economic incentives to decrease waste water effluents, these retributive taxes must be applied in proportion to the volume of polluting substances discharged. Otherwise, the rates would become a simple revenue-raising mechanism for the environmental authorities, and would have no major incentive effect on the polluters to reduce the negative impacts of their discharges.

However, the tax can not cover pollution from non-point sources such as agricultural and urban run-off, since measuring the quantity of contaminants contributed by each source would be an impossible task. Therefore, the new tax system will be restricted to those industries and residential locations operating as point sources of pollution and to sanitation companies handling discharges from these sources.

Tax Level Determination

Theoretically, it would seem advisable not to establish limits to individual discharges, in order to give this instrument the maximum possible flexibility and to allow for gains in efficiency. As already mentioned, the retributive taxes are expressed in the present legislation as a typical pigouvian tax and not as a mechanism to generate financial resources for the environmental authorities. However, the practical feasibility of applying these taxes remains a question.

According to Von Ameberg (1995), the definition of retributive tax in the Colombian legislation constitutes 'the most explicit example of a pigouvian tax expressed within a law', although he doubts its applicability since 'the development of the economic evaluation of the environmental damages is very rudimentary as of now'. The process followed in Colombia to tackle this problem consists of the adoption of intermediate tax levels, not set according to the optimum determination of discharges (equalling costs and benefits at the margin) but rather aimed at reaching previously-determined environmental targets.

Law 99 outlines the series of steps that should be taken to define the water pollution charge levels. These are described below.[5]

1   Identification of the previously established environmental goal (ie maximum accepted level of concentration of pollutant substances in a water source).
2   Identification of reduction and abatement measures required for the total levels of discharges to meet the pre-established goal.
3   Estimation of pollution abatement costs based on different tax levels. This estimation should take into account the technological options available for pollution abatement.
4   Definition of tax levels to reach the pre-defined environmental goal.

In general, these procedures still preserve the pigouvian concept established by the law but apply it in a gradual manner. In the first stage, an effort is made to achieve the environmental goals at minimum cost. In the long term, with the application of valuation techniques to put a value on environmental damages, rates will get closer to the optimum levels defined by law. This charge-setting would fall under the responsibility of local authorities, taking into account the 'diversity of regions, availability of resources, their assimilation capability; the polluters involved; the socio-economic conditions of the affected population; and the opportunity cost of the resource involved' (Law 99/1993).

Simulation of the Economic Impact of Water Pollution Taxes

This section presents some of the results produced by applying a General Equilibrium Model for Colombia to simulate the economic impact that the application of the water pollution taxes (retributive rates) could have on the manufacturing industry. The effects of the taxes are analysed at both the macroeconomic and sectoral levels.

The estimated water pollution taxes were taken from of a recent study (Ferreira and Roda, 1996) which estimated the rates that would be required to reach different environmental goals (expressed as pollution control grades). This study assessed the unit cost of control for different types of pollutants based on a sample of 60 Colombian industrial companies. Based on this cost function, the corresponding control levels of 30 per cent, 60 per cent and 90 per cent were assessed for each of the water pollutants. These control level percentages refer to the extent to which industry would be required to decrease the levels of pollutants in discharge waters: by 30 per cent, by 60 per cent or by 90 per cent of the original levels. The pollutants on which these calculations are based are BOD, COD and TSS.

Table 17.1 presents the expected impact of the water pollution taxes on different sectors of the country's economy. The relative impacts on the different sectors vary with the control levels applied. For example, at a 30 per cent level of control, the highest increase occurs in the paper and printing sector, which would have to increase total tax payment by 1.45 per cent. At a 60 per cent control level, other industries would be significantly affected, particularly the meat processing industry (including slaughterhouses) with a 11.35 per cent increase in total tax payment, and the chemical industry with an increase of 7.92 per cent.

It must be observed that with the exception of the paper and printing industry, at the 30 per cent control level, the increase in total tax cost would be below 1 per cent for all sectors. Even for tax rates at a 60 per cent control level, the increases in total tax costs are above 1 per cent for only meat (1.94 per cent), chemicals (1.35 per cent) and processed foods (1.24 per cent).

Table 17.1 *Simulation of Water Pollution in Sectoral Costs in Colombia*
(tax costs as percentage of GDP)

Sector	Control level		
	30%	60%	90%
Foods	0.59	1.24	7.30
Meat Processing Ind.	0.93	1.94	11.35
Beverages	0.10	0.20	1.15
Textiles and Leather	0.31	0.65	3.79
Wood and Furniture	0.32	0.67	3.92
Paper and Printers	1.45	3.03	17.79
Chemicals	0.65	1.35	7.92
Non-Metallic Minerals	0.17	0.32	1.76
Metal Products	0.06	0.13	0.76

The main macroeconomic impacts of the retributive water tax are analysed in Table 17.2. A tax rate corresponding to a 90 per cent control level would result in a 0.04 per cent drop in GDP. If greater resources are returned to the economy through the expansion of environmental public expenditure, the drop in the GDP would only be 0.017 per cent. These 'compensatory investments' would be possible if a compensatory component is added to the retributive component of the tax.

Table 17.2 *Estimated Macroeconomic Effects of Water Pollution*
Taxes in Colombia

	30%	60%	90%	With compensatory investments
GDP	−0.003	−0.007	−0.040	−0.017
Consumption	−0.005	−0.011	−0.067	−0.046
Exports	0.000	−0.002	−0.010	−0.014
Imports	−0.002	−0.005	−0.029	−0.005
GDP Deflator	0.001	0.003	0.016	0.021
Consumer Price Index	0.002	0.003	0.019	0.025
External Savings	0.000	−0.001	−0.005	−0.002
Fiscal Surplus	0.001	0.002	0.013	0.005
Private Savings	0.000	−0.001	−0.004	−0.003
Investment	0.000	0.001	0.004	0.000
Urban Employment	−0.003	−0.006	−0.039	−0.016

At a 90 per cent control level, the taxes would generate an additional 0.019 per cent increase in the consumer price index. The scenario with public compensatory investment would accentuate this trend, increasing inflationary pressures by 0.025 per cent. In any case, the magnitude of the inflationary impact of the pollution taxes is minimal. As a comparison, a 20 per cent increase in the price of gasoline would have an inflationary impact of 0.33 per cent without public compensatory investment (which assumes a drop of almost 1 per cent in the GDP). A 20 per cent devaluation accelerates the inflation rate by more than four percentile points. Therefore, the inflationary impact of the water pollution taxes, even at a 90 per cent control level, is very low.

The taxes' impact on growth is also very low. Compared to the drop of almost 1 per cent in GDP from a hypothetical 20 per cent increase in gasoline prices, the same percentage increase in water pollution taxes would lower GDP by only 0.017 per cent (with compensatory investment). This drop in GDP may be even lower, since companies would prefer to control their emissions rather than pay the increased rates. The impact of the water pollution taxes on the companies' costs and on their economic activity would therefore be less than these figures suggest.

The model also estimates the taxes' impacts on price and production levels. For a 90 per cent control level, a major impact is observed in production in the following sectors: beverages (with a drop of 0.21 per cent in gross production), paper and printing (with a 0.10 per cent decrease) and other agriculture/livestock food. The same sectors would also show the most significant price increases, reflecting the pressure on costs as a result of rate imposition. In the case of beverages, prices would increase by 0.21 per cent.

Changes in the price of industrial goods reflect two opposite trends. In some cases the taxes' impact on costs would dominate, which in turn would result in increased prices. In other cases the taxes' recessionary effect (although slight) would prevail, resulting in a reduction in economic activity and a decrease in prices.

On the whole, the results of the model suggest that the impact of applying water pollution taxes is very slight for the industrial sector, even at 90 per cent control levels. Considering these results and the positive environmental effects which would be expected (which were not quantified by the model), these simulations suggest that the economic viability of implementing water pollution taxes is high. Applying a model which would allow enterprises to adopt pollution control technologies, rather than pay increased taxes, would strengthen this conclusion even more.

CONCLUSIONS AND RECOMMENDATIONS

Setting an Environmental Goal

As far as water resources are concerned, an environmental goal can be defined in terms of the acceptable concentration of the contaminant substance in a water body. It is also possible to define an environmental goal in terms of specific water uses for a given watershed. These definitions must be undertaken by the environmental authority. However, in order to ensure consensus building and information sharing, this process should be based on a wide consultation with all stakeholders in the affected area.[6]

Water users' participation not only increases the practical feasibility of the policy, but also may partially replace the need for an economic valuation of environmental resources and costs. An agreement over environmental goals may reflect the value a community places on damages from pollution. Once the concentration goal for pollutants is identified, it is necessary to know the aggregate level of discharges that are actually taking place and their effects on the quality of the receiving body. In this respect, a discharge recording process has been initiated in Colombia, which will need to be consolidated and extended over the whole country, or at least to the most significantly polluted areas. The efficiency and applicability of this pollution tax system in the future will therefore depend on two main efforts: (i) increasing the technical capacity of the rural/urban environmental authorities; and (ii) improving the environmental information on discharges and sources of water pollution.

Administrative Aspects

One of the most complex aspects associated with the operation of retributive taxes concerns the measurement of the contaminant discharges to assess the appropriate tax levels. In this respect there are useful examples to draw on in Colombia. In cities such as Bogotá, Medellin, Barranquilla and Cartagena, certain monitoring efforts have been undertaken. The city of Cali is an exceptional case. As a result of the prior application of retributive rates, there are systematic records available over a period of several years for a significant number of firms.

It has also been proposed that companies should undertake self-monitoring, which would then be subject to random checks, to avoid the need to build up a sophisticated infrastructure within the environmental agencies.

For cases where firms' sewage is collected and discharged through the general sewage system run by sanitation companies, taxes are to be levied on the sewage companies. Sanitation companies can then transfer these additional costs to users by increasing tariffs. However, in order to discourage pollution, this increase in tariffs must be proportional to the discharge of each original source.

Sanitation companies will also be encouraged to reduce their tax costs to the point where the marginal cost of such reduction is equal to the marginal control cost. In conclusion, with this type of mechanism, not only are industrial companies motivated to control their discharges but sanitation companies are also encouraged to invest in water pollution control plants.

NOTES

1   Instituto de Pesquisa Economica Aplicada, Brazil.

2   FEDESARROLLO, Colombia.

3   FEDESARROLLO, Colombia.

4   For a brief explanation of a pigouvian tax, see glossary.

5   See Ministry of the Environment (1996).

6   In Law 99/1993, the participation of non-governmental organizations and other members of civil society on the board of directors of the regional EPAs is mentioned as an important mechanism of citizen participation in environmental policy making.

REFERENCES

Ferreira, P and Roda, J P (1996) Tasas por Contaminación Industrial del Agua. *Revista Planeacion y Desarrollo, Deparamento Nacional de Planeacion*, vol 17 (2), Bogotá.

Rodriguez Becerra, M and Uribe, E (1995) Instrumentos econômicos para la gestión ambiental en Colombia, CEPAL, mimeo, Bogotá.

Ministry of the Environment (1996) Programa para la regulamentacion de las tasas por vertimientos líquidos, Comisión de Tasas Retributivas, Documento de Trabajo, Bogotá.

Von Ameberg, J (1995) Selected experiences with the use of economic instruments for pollution control in non OECD countries; in: Borregar, N et al (eds) Uso de Instrumentos Econômicos en la Política Ambiental: Análisis de Casos para una Gestión Eficiente de la Contaminación en Chile. CONAMA, Santiago.

18

AIR POLLUTION TRADABLE
PERMITS IN SANTIAGO, CHILE

Ronaldo Seroa da Motta[1] *and*
Monica Rios Behrem[2]

BACKGROUND

Air pollution in Santiago is a serious problem. Every winter, the ambient air quality standards are violated for suspended particulate matter (PM-10) and carbon monoxide (CO) while in the summer excessive levels of ozone (O_3) are the norm. One of the major consequences of this situation is the health cost associated with air pollution, which is estimated to be around US$100 million.[3]

Mobile emissions (ie emissions from vehicles) consist of mainly PM–10, nitrous oxide (NO_x) and CO, while industrial fixed source emissions are largely SO_2 and PM–10. Group sources, such as street dust and residential wood burning, pollute mainly with PM-10 but their contribution to air quality is not significant enough to affect overall concentration levels. Katz et al (1994) estimated that mobile sources are the major cause of air pollution in areas where concentrations exceed ambient standards.

Santiago's air pollution problem became evident during the 1980s though no effective action was taken until the beginning of the next decade, when the Aylwin government took office in 1990. Previous studies had already evaluated the health effects of the city's air pollution back in 1986 and confirmed that air pollution by ozone and particulates violated the ambient quality standards set in 1978 by the Ministry of Health.

The Special Commission for the Decontamination of the Metropolitan Region (CEDRM) was created in 1990 and prepared a 'Master Plan', covering air and water pollution and solid waste management. The plan included several proposals for controlling air pollution from mobile and fixed sources, and these were then implemented during the subsequent years. The CEDRM used an air dispersion model, which assumed a linear relationship between total emissions and pollutant concentration.

The Aylwin administration (1990–94) maintained a free market economy approach towards economic policy and did not impose any technological requirements for pollution control on the dynamic productive sector. The administration introduced an economic instrument known as 'tradable permits' to encourage air quality improvements by firms. This instrument operates alongside the emission standards regulations and functions in the following manner. Existing enterprises are allocated permits, free of charge, according to baseline figures for maximum permissible levels of different pollutants. These enterprises can then choose to reduce their air pollution below the baseline levels, have this reduction recognized, and then sell their permits to other firms wishing to expand or move into the area.

Negotiations between the Aylwin government and the private sector began once this instrument had already been devised, and the talks focused largely on the level of emission standards, the enforcement mechanisms and the gradual nature of the instrument's application.

The General Law on the Environment (March 1994) set guiding principles and goals for Chilean environmental policy. This Law defines a system of environmental impact assessments, the setting of standards, prevention and decontamination plans, and the creation of the National Environmental Council (CONAMA) as the environmental agency in charge of co-ordinating environmental policies. According to the General Law on the Environment, tradable permits are considered one of the alternative instruments for regulating environmental problems such as air and water pollution and solid waste management.

National air quality standards have now been established by the Ministry of Health for each of the most important pollutants. These standards are not legally binding, but are rather goals to be aimed at when making new policies. Major regulation initiatives for mobile sources include the mandatory retirement of old buses; the introduction of vehicle emission standards, which in practice makes the use of 3-way catalytic converters mandatory; limits on the operation of vehicles based on the licence plate number for non catalytic engines; and the auctioning of routes for urban buses in Santiago.

Regulations for fixed sources include a ban on the use of wood-burning fireplaces during the entire year; the setting of emission standards; and decontamination plans for some fixed sources in Santiago. A system of tradable permits for PM-10 emissions from industrial sources has also been established.

Most of the environmental regulations in Chile are based on a strategy of 'Command and Control' – ie the traditional approach of directly regulating activities which affect the environment, rather than using market mechanisms to indirectly influence these activities. A previous experience with tradable water rights in 1981, based on water rights dating back to the beginning of the century, does however confirm Chile's predisposition to market solutions. Nonetheless, it is the current commitment to market forces in the whole Chilean economy that has been the main driving force behind tradable permits in environmental management. In this market-oriented policy environment, taxation is usually regarded as a second best solution.

LEGAL CONTEXT

The system of tradable air quality permits in Santiago is legally defined by Supreme Decree No 4 (SD4)–1991, from the Environmental Health Service of the Ministry of Health, which states that:

- Fixed sources must comply with a uniform emission concentration standard and offset their PM-10 emissions by trading permits with each other.
- Fixed source emissions of particulate matter under 1000 m^3/hr are categorized as 'group sources' while fixed source emissions over 1000 m^3/hr are termed 'stationary sources'.
- Existing stationary sources were required to comply with an emission standard of 112 mg/m^3 before the end of 1992 and to offset any emissions in excess of 56 mg/m^3. This emission standard had been established by the Ministry of Health in 1978, but was not enforced until recently.
- Group sources were required to comply with the standard only, without participating in emission trading.
- New sources were required to comply with an emission standard of 56 mg/m^3 before starting their operations and to offset 100 per cent of their emissions, according to the following schedule: at least 25 per cent of total emissions by December 31, 1993; 50 per cent by December 31, 1994, reaching 100 per cent by December 31, 1996.

The initial distribution of permits to existing sources was based on the pollutant gas flows reported in 1992, multiplied by an emission concentration standard of 56 mg/m^3, assuming 24 hours of operation, regardless of the actual operating times. The initial permits were distributed free of charge. New sources do not receive any permits and must obtain their permits from existing sources.

The Supreme Decree No 812 of 1995 (SD 4), was a supplementary decree outlining the procedure for implementing the offset system. It requires permit holders to present an application for emission offsets together with an annual emission report with emission levels by sources.

Table 18.1 summarizes the requirements for both group and stationary sources.

Table 18.1 *Particulate Matter Emission Standards and Deadlines for Fixed Sources*

Type of Fixed Source	Jan–1, 1993– Dec 31, 1997	Jan 1, 1998– onwards	Offset Criteria
Stationary source: over 1.000 m³/hr			
Existing Sources	112 mg/m³	56 mg/m³	with offset of excess
New Sources*	56 mg/m³	56 mg/m³	with 100%offset
Group Source: below 1.000m³/hr			
Existing Sources	112 mg/m³	56 mg/m³	without offset
New Sources*	56 mg/m³	56 mg/m³	without offset

Note: *According to SD 4, the expansion of existing sources is legally treated as a new source

The Control Fixed Sources Programme (PROCEF), created in 1993 with an initial investment of US$1 million and recently linked to the Ministry of Health, is responsible for the implementation and enforcement of Supreme Decrees No 4 and No 812. The National Environmental Council (CONAMA), the national executive environmental agency, is currently working on the design of a law for tradable permits for various environmental problems, which will eventually replace both these decrees.

The calculated investment in pollution control equipment in Santiago is around US$50 million, assuming abatement investments have taken place in half of the 1334 fixed sources in the system area. Other costs borne by the private sector include those associated with the reporting of emissions, as enterprises need to contract the services of private laboratories to perform isokinetic measurements.

Commenting on the economic impact of the emissions trading system, O'Ryan (1995) states 'costs savings in Santiago will be in the tens of thousands of dollars for only a very few firms. The incentives for trading will be low, and it can be expected that only large sources will engage actively in trading'. This conclusion sounds reasonable since this type of permit market exhibits high transaction costs.

IMPLEMENTATION ISSUES

The effectiveness of this tradable permit system for air pollution control is hard to assess since the system has not yet been fully implemented. Only a few applications for offsets have been presented to PROCEF, all of which correspond to expansions of existing firms.

The improvement of air quality in Santiago in the last five years cannot be isolated from other measures undertaken to control pollution from public transport and other mobile sources.

According to Katz et al (1994), the percentage of fixed sources not complying with the emission standard (112 mg/m³) decreased from

December 1992 to December 1993 from 16.6 per cent to 4.6 per cent of all sources. The daily emissions decreased from 15.3 to 8.1 metric tons over the same period. The authors conclude that the air quality control measures had a significant environmental impact, but that this can be attributed to compliance with the regulations (ie the Command and Control component) rather than participation in the tradable permits component.

PROCEF has encountered considerable difficulties in implementing the air pollution control system. Firstly, it had to build from scratch its own institutional capacities for pollution control and, even though PROCEF has increased its monitoring capacity over time, it still needs more financial resources to be able to improve its work. To fully implement the system, PROCEF's budget would have to be expanded from the initial US$1 million to around US$7–8 million per year.

Secondly, the initial inventory data on emissions were found to be significantly lower than the actual emissions occurring. PROCEF had to spend many months developing a comprehensive and detailed registration of point sources, their levels of emission and concentrations. The quality of their information has improved substantially, particularly in the case of boilers and heaters. However, data problems still exist for other polluting industrial processes.

Given this situation, emissions offsets are not yet required for these industrial sources and compliance deadlines have been postponed. Moreover, once new inventory data become available, substantial increases in the initial permit allocations are expected, which will induce firms to postpone any trading of permits.[4]

The private laboratories, which are contracted by firms to measure emissions, have until recently been the source of some implementation problems, including corruption and the use of inadequate equipment. This situation has been improved by PROCEF requiring the certification of the laboratories' measuring equipment, the training of their personnel, and other measures to control these laboratories.

The PROCEF programme was recently transferred to the Ministry of Health, which traditionally has adopted a more regulation-oriented approach rather than a market-based one. This partially explains why the implementation of the offset system has been left behind. More and more sources are complying with the emission standards, but less priority has been given to the application for offsets and the compliance with offset deadlines.

CONCLUSIONS AND RECOMMENDATIONS

There are some other issues that should be considered when analysing the offset system of air pollution in Chile, such as: the design of the system; its spatial dimension and the transaction costs involved.

The design of the system does seem to require modification, to account for the fact that particulate matter (PM) of different size and toxicity has a different impact on air quality. PM of different toxicity should be 'weighted' differently for the purpose of an offset transaction, or one could end up having lower PM emissions but of higher toxicity or carcinogenic potential. The other problem is related to the fact that ambient air quality standards are currently defined in terms of PM-10, while emissions at the chimney level are defined in terms of total PM.[5]

Another weakness of the system relates to its spatial dimension. The SD 4 does not impose any restrictions on the location of the sources participating in the offset transaction, although emissions from one trading source may generate quite a different impact on ambient standards than similar emissions in a different location.

Markets for the tradable permits are less dynamic than desired because of the existence of transaction costs, such as search and information costs, and bargaining and decision costs. Since there is no organized market for trading permits under the current system, one would expect more trading within expanding firms because it reduces the transaction costs of finding a trading partner. In fact, all sources that have requested recognition of emissions are planning to offset with these types of intra-firm transactions (Katz et al, 1994; O'Ryan, 1995). Given the existence of transaction costs, it will probably be the large firms who can overcome them and obtain positive benefits from offsetting their emissions. Brokerage and information sharing will be required to help encourage inter-firm trade.

It must be recognized that the system has yet to be tested and this will only occur when full implementation gets underway. The strict standards required at the pre-trade phase have induced existing firms to comply and has reduced their willingness to trade. That disincentive will be less significant to new firms wishing to install plants in the system area, and one could then expect a large interfirm trade movement.

Finally, it is worth mentioning that the full implementation of tradable permits for air pollution in Santiago will largely depend on the strength of institutional capacity to assure monitoring and fiscalization for efficient permit allocation and trade.

NOTES

1   Instituto de Pesquisa Economica Aplicada, Brazil.

2   Comision Chilena del Cobre, Chile.

3   According to World Bank (1994).

4   O'Ryan estimated an increase of almost 75 per cent in some cases.

5   PM-10 refers to particulate matter measuring less than 10 microns in diameter, while total PM refers to total particulate matter emitted, including particles with a diameter of more than 10 microns.

REFERENCES

CONAMA/University of Chile, Deparemento Ingeniería Industrial (1995) Diseño del Sistema de Permisos de Emision Transables, Estudio 0001–002/94, Informe Final.

Giaconi, J M S (1993) Consideraciones de Toxicidad en el Control de Emisiones a la Atmósfera; in: Antecedentes para la Generación de un Sistema de Derechos de Emisión Transables. Una Aplicación a la Contaminación del Aire. Comisión del Medio Ambiente del Centro de Estudios Públicos, Documento de Trabajo No 207.

Hartje,V, Gauer,K and Urquiza, A (1995) Uso de Instrumentos Economicos en la Politica Ambiental Chilena; in: Borregaard, N et al (eds) Uso de Instrumentos Economicos en la Politica Ambiental: Analisis de Casos Para una Gestion Eficiente de la Contaminacion en Chile, CONAMA, Santiago.

Katz, R (1993) Compensación de Emisiones: Un Instrumento de Alcance Global para el Control de la Contaminación Atmosférica; in: Antecedentes para la Generación de un Sistema de Derechos de Emisión Transables: Una Aplicación a la Contaminación del Aire. CONAMA, Centro de Estudios Públicos, Documento de Trabajo No 207.

Katz, R, Sanchez, J S and Balcazar, R A (1994) Analysis of an Emission Offset System for Particulate Matter in Santiago, Chile. Draft version.

O'Ryan, R (1995) Emissions Trading in Santiago: Origins, Current Situation and Some Lessons. Paper presented at the Third Annual World Bank Conference on Environmentally Sustainable Development, Washington, DC.

Rios Behrem, M and Quiroz, J (1995) The Market for Water Rights in Chile: Major Issues. World Bank Technical Paper 285, The World Bank, Washington DC.

Stavins, R N (1995) Transaction Costs and Tradeable Permits. *Journal of Environmental Economics and Management*: 133–148.

World Bank Chile (1994) Managing Environmental Problems: Economic Analysis of Selected Issues. Report No 13061-CH, Environment and Urban Development Division, Country Department I, Latin America and the Caribbean Region, The World Bank, Washington, DC.

19

FORESTRY TAXES AND FISCAL COMPENSATION IN BRAZIL

Ronaldo Seroa da Motta[1]

BACKGROUND

Forest conservation in Brazil is legally established in the Forest Code and water legislation, largely by these regulations placing restrictions on land-use within forested areas. The Forest Code, published in 1934 and revised in 1965, regulates the use of wood from forests, defines conservation units, restricts farming and logging activities, sets criteria for burning and chainsaw uses and defines a system of sanctions and fines.[2] Legislation on water use has also introduced very strict rules for land use in forest areas, largely to protect the watersheds for domestic water supply purposes.

In spite of the severity of these regulations, large areas of Brazil's forests have been cleared. Less than 8 per cent of the Atlantic forest, previously found in the most developed areas of the country, remains. The savannah areas of Cerrados, in the central region, have already been cleared to half their original size, for farming purposes. Although less than 10 per cent of the Amazon forest has been cleared, annual rates of deforestation are very high at around 0.3 per cent.[3]

Apart from problems of institutional weakness to enforce the forest conservation regulations, other factors have played a role in the deforestation of important ecosystems in Brazil, eg a highly concentrated land tenure system, where small farms (of less than 10 ha) cover less than 3 per cent of the total farming area while large farms (with more than 10,000 ha) cover 40 per cent. Additionally, very low productivity levels encourage continuous land reclamation. Personal income is also highly concentrated with 66.1 per cent of total income accruing to the richest 20 per cent of families, while just 2.3 per cent accrues to the poorest 20 per cent.[4] Such inequality creates an immense surplus of low-income workers ready to seek work in frontier areas.

Credit and fiscal systems favouring agricultural activities have been put in place without any regard for the agro-ecological conditions or managerial practices of the farming enterprises in the areas concerned. Land reclamation and land taxation rules are based on land use criteria such as the amount of area allocated to farming, thereby encouraging and legalizing clearing. Due to the high wood values in these frontier areas, logging activities play an important role in financing clearing or even taking advantage of legal clearing licensing.[5] And finally, regional development programmes in frontier areas based on road construction, although mostly phased out, have contributed immensely to migration flows and economic land use activities.

Some of these factors can not be easily reversed as they would require long-term structural adjustments to alleviate social inequalities, to achieve satisfactory land reform and to solve the financial problems which limit the capacity of governmental agencies. However, there is still the possibility of applying economic incentives to mitigate the current trend towards deforestation and biodiversity loss.

Three important economic instruments have been applied in Brazil to try and control deforestation. Two are forestry taxes and the third is fiscal compensation through conventional taxation.

In the frontier areas of Amazonia and Cerrados, the application of taxes is very difficult to enforce, due to their vast size, the lack of infra-structure, and very low population densities. Therefore, one might expect that such instruments can play only a limited role in creating market-based mechanisms for biodiversity control, though they can be powerful revenue-raising instruments to help strengthen institutional capacity. Fiscal compensation, on the other hand, has a very low admin-istration cost and creates a real incentive for those engaged in conservation measures.

A brief description of the use of these instruments in Brazil is presented here, along with some recommendations for their revision.

THE CASE OF FORESTRY TAXES IN BRAZIL

The National Reforestation Fund in Brazil

The Brazilian Forestry Code states that those exploiting, using, trans-forming or consuming raw material from forests are obliged to undertake reforestation of appropriate species, equivalent to their consumption level. This requirement covers logging as well as the consumption of charcoal and firewood of unknown origin.

Since 1978, however, a federal norm gives those users who consume less than 12,000 m³ of forest raw materials per year the option to pay a deforestation contribution instead of investing in reforestation.[6] The rationale for this measure is based on the assumption that reforestation by small consumers is very costly to undertake and monitor, as there are no benefits from economies of scale. Therefore, a government fund composed of this deforestation contribution would generate revenue, which could be used to plan a more efficient reforestation, taking into account diverse social and political concerns.

This reforestation tax was therefore an attempt to create a national fund for reforestation, which would make forest users comply with the legislation without disrupting their own activities. However, the tax level was fixed without due regard to reforestation cost variations and the funds generated were never actually used for reforestation. As will be discussed later, these implementation failures were also exacerbated by the weak capacity of the institution responsible for administering this tax.

Implementation Issues
The low level of the contribution required seems to be the main reason behind the failure of this instrument to change forestry production patterns in Brazil.

This tax, even if designed for revenue-raising purposes, could also have encouraged reforestation measures, as long as the tax rate was high enough. But that was not the case and all those eligible to pay the tax (ie those consuming less than the legal limit) have opted to pay it rather than invest in reforestation themselves.

Moreover, the value was fixed at a level of approximately US$4.00 per m^3 of wood, with no variation between species beyond a distinction between conifers and non-conifers. Although indexed to inflation, the real value of the tax has not taken into account the escalating costs of reforestation.

Finally, funds from this contribution have been used largely for IBAMA's own budgetary needs, rather than for actual reforestation activities, and this may explain the complete lack of revenue records available to the public. Only recently has the government allowed part of this revenue to be diverted to states and NGOs willing to invest in forestry activities in municipalities where reforestation can either create economic opportunities or rehabilitate deforested areas.

Conclusions
The institutional fragility of IBAMA to effectively carry out its legal responsibility for collection of the tax has been a major barrier to the introduction of any kind of fiscal system within the forest policy in Brazil. Considering the size of the country and its forests, IBAMA should have very strong political and institutional powers to undertake its mandate. However, the public sector crisis in Brazil, reinforced by tighter control on public expenditures, has had a detrimental effect on the environmental agencies.

IBAMA has been facing a long crisis from restricted budgets and lack of human resources *vis à vis* a growing demand for environmental management in the country. The implementation of this fiscal device was therefore designed to be as simple as possible, to reduce the administrative burden on IBAMA. The revenue was therefore easily captured within IBAMA as an important source of income, rather than as a fund for financing reforestation.

If a tax or user charge is designed to curb deforestation, the tax level should be such that reforestation should be less costly than the tax or charge payment, to induce forest exploiters and consumers to undertake reforestation. Tax and charge levels should also take into account differences in scarcity and reforestation cost between different ecosystems and species.

Forestry Tax in Minas Gerais State in Brazil

Since 1975 the state of Minas Gerais has attempted to introduce a forestry tax to finance the monitoring and enforcement activities of the state Forest Institute.

The state of Minas Gerais, with its large reserves of iron ore, is the biggest steel and pig iron producer in the country. The pig iron and steel industries, present in the state since the beginning of the 1960s and still expanding, use charcoal as their main energy source. The forestry tax was a response by the environmental agencies to try and support their work on monitoring and enforcing the forest legislation.

Taxation is levied on all forest products – from logs and firewood to roots and seeds – consumed or transformed in economic activities. This tax is, in fact, a kind of user charge similar to the national tax discussed above, although in this case the aim is *explicitly* stated as financing the environmental agency rather than funding reforestation. Taxes are also levied in the case of legal deforestation for agricultural purposes. The tax value was defined at 3 per cent of the value of forest products and collected by the state treasury.

For almost 15 years the tax was subject to a judicial dispute between legislators and tax-payers, the latter group arguing that the state value added tax was supposed to fulfil any budgetary need and, therefore, the forest tax was a double tax. The outcome of this dispute was a change in the law, introducing a forestry tax level based on percentages of an indexed currency varying according to each type of forest product. Reductions of up to 50 per cent of the tax due can now be granted to those undertaking reforestation which will generate forest production equivalent to the consumption level.

Since the full implementation of this tax got underway in 1992, it has been a key factor in changing the pattern of charcoal consumption in the state.

Implementation Issues
The outcome of the judicial dispute was an unexpected one, since it has allowed tax differentiation, and has therefore turned the tax into a very effective economic incentive to curb charcoal production. Today, this fiscal device can be considered a deforestation tax as it differentiates between species and products and allows the Forest Institute to penalize certain uses by altering the percentages.

These adjustments have been made and in December 1993 a revised set of tax levels was published. These showed that users of charcoal and firewood from native forests (an important source of deforestation in the state) were charged four to five times as much as in the previous list, whereas taxes on other uses have increased by no more than 100 per cent. Records on the tax application are still not well organized or fully available but it is estimated that a revenue of US$11 million was collected in 1993.[7]

The revenue generated from this tax was a key factor in enhancing institutional capacity of the Forest Institute in its various locations within the state. Such strengthening has improved monitoring performance and tax revenue collection.

Although it is very early to assess, the current pattern of wood consumption in the state seems to be changing too. For example, the share of wood supply from native forests in total charcoal and firewood production has declined from 70 per cent in the 1980s to less than 50 per cent in recent years.

The total environmental effects are, however, very difficult to determine. If, on the one hand, an increase in reforestation initiatives has been noticed, on the other hand it is clear that part of the state demand for wood has been met by supplies from neighbouring states where such heavy taxes are not being applied. Such supply deviation, apart from resulting in the inevitable loss of forest resources in the supply region, has promoted rapid urbanization of remote areas without adequate infrastructure. The Minas Gerais forestry tax can therefore be viewed as a typical case of distortionary taxation with respect to spatial resource uses.

Conclusions
It is worth noting that this tax was primarily conceived as a cost-recovery instrument. Its use as an economic incentive depends on the political will prevailing in the state. However, the determination of tax levels is still made on an ad hoc basis, without detailed modelling of the taxpayers' responses. The effectiveness of this tax as an instrument to change behaviour may still face strong political barriers and other judicial disputes as it is legally aimed at revenue generation.

Today the charcoal sector and other industries (such as metallurgy and cement producers) who rely on charcoal, are already seeking financial help from the state and national treasury to counteract the impacts of the higher costs they face by either paying this tax or finding other raw material sources. The objective of forcing a phase-out of these activities may not succeed on political grounds since the pig iron and steel sectors are important sources of employment and fiscal revenue in Minas Gerais.

The conversion of this tax into an effective economic instrument, to sustain the current improvements in forestry activities, may require new legislation and a political compromise. And clearly, to fully achieve the anticipated environmental benefits, this tax has to be jointly adopted by neighbouring states to avoid transboundary supply deviation.

FISCAL COMPENSATION FOR LAND-USE RESTRICTIONS IN BRAZIL

Legal Context

Revenue from valued added tax (ICMS) in Brazil is collected by states and 25 per cent is distributed among municipalities. The amount of revenue distributed by the states to the different municipalities is defined by the Constitution. At least 75 per cent of the municipalities' share has to be distributed according to municipality participation in total fiscal revenue generation, and the other 25 per cent share can be distributed according to any criteria set by the State Assemblies. Since 1992, three Brazilian states – São Paulo, Rio de Janeiro and Paraná – have introduced an ecological revenue distribution criterion for this 25 per cent share, based on the area of land in each municipality which is subject to land-use restrictions for conservation and watershed protection.

The main aim of introducing this criterion was to create a budgetary increment, without the need of a new fiscal instrument, to compensate municipalities where considerable land-use restrictions were in place. It was recognized that these restrictions, while generating ecological benefits for society as a whole, could impose barriers to economic development in these regions and could consequently reduce their income level. Moreover, it was expected that these additional resources could also encourage other municipalities to implement sustainable activities.

The criterion is applied with a scoring system that reflects the acreage of effectively protected land in the municipality. This required the development of several indicators to measure protection performance in the entitled municipalities.

The introduction of this environmental criterion was conducted with due consideration of the political context, and involved mayors and representatives in the state Congress. Since the total state tax revenue was not being changed, the new criterion reduced the share of other municipalities and so political resistance had to be acknowledged.

As a first step, it was agreed that a very small percentage, varying from 2 per cent to 5 per cent of the 25 per cent share whose disbursement can be determined by the states (equivalent to between 0.5 and 1.25 per cent of total ICMS revenue), would be devoted to this ecological criterion. Nevertheless, legislation allows for periodical revisions of the criteria, indicators and revenue share, based on the results seen. These revisions have in fact occurred in São Paulo and Paraná, where the system has been in place since 1992. Major changes announced so far have been limited to the redesign of indicators to help evaluate protection performance.

Implementation Issues

Since this instrument is based on conventional taxation, its collection and distribution have been easily undertaken. The main implementation problem relates to the definition and determination of indicators to calculate the different municipalities' shares.

The first step in determining each municipality's share is to set conversion factors for each type of conservation unit and watershed area, according to their respective ecological and social importance. Using these factors one can then calculate the total area entitled to green ICMS revenue.

Secondly, the state Environmental Protection Agency (EPA) defines criteria for area protection performance based on a performance index developed for each entitled area. The weighting sum of area size and performance indicators determines each municipality's share.

Calculating the conservation and watershed areas is usually a simple process, based on legal documents, and does not rely on the institutional capacity of the EPAs. However, estimating the performance indicator relies solely on the EPAs' capacities and requires significant field research and monitoring. Moreover, the performance indicator criterion is not free of political dispute and tends to be weakly implemented to try and avoid disputes.

In Paraná, since 1992, 5 per cent of the ICMS revenue is diverted to 112 municipalities where there is restricted land-use – 2.5 per cent going to ecosystem conservation areas and another 2.5 per cent to areas for water supply sources protection. The total revenue accruing to municipalities based on these preservation criteria has been around US$50 million per year. Distribution of this compensation allowance to municipalities is fixed according to the importance of the protected area, based on the size of the area and the degree of restriction stated in the legislation which created them. Paraná's EPA then evaluates the compliance of these municipalities with the required environmental quality standards operating in these areas to determine the amount to be paid.

This monitoring system is still under implementation but it has already encouraged several municipalities to consider other activities more appropriate to their natural endowment, such as ecotourism. This source of revenue has also helped promote investment in sanitation and other urban infrastructure services.

In São Paulo, similar legislation was approved at the end of 1993 for areas of ecosystem protection and hydroelectricity generation, with a more modest fiscal compensation equivalent to 2 per cent of the 25 per cent share of ICMS revenue distributed to municipalities according to the state's discretion. Revenue distributed according to such criteria in the State of São Paulo in the 1994/1995 period was around US$19 million. Since it is a very recent instrument and still small in fiscal magnitude, it is difficult to predict the resultant impacts on municipalities. However, another law related to conservation areas is under discussion in the state legislature and it is expected to offer higher incentive levels.

In the Rio de Janeiro state, similar legislation was tabled in 1993 to be approved by the state legislature but has not been enforced due to political reasons. The initial percentage of 8 per cent proposed in the law project was reduced to 3 per cent and it will be deducted from the ICMS revenue due to municipalities. The distribution criteria would assist conservation areas of ecosystems and water supply sources as well as the implementation of environmental planning in these municipalities. The expected revenue under this distribution criterion is more than US$50 million.

Conclusions

Although monitoring costs can be high, the 'green value-added tax revenue' of Brazil can be seen as a good example of a low collection cost instrument where political barriers were carefully removed and existing fiscal legislation was used. Its objective of compensating ecological services has an important effect in achieving social equilibrium and creating incentives for conservation.

However, the criterion for performance indicators needs to be well defined and the EPAs have to be institutionally strong to properly apply them. Without the proper application of performance indicators, incentives for conservation will fail as beneficiaries will not perceive changes in their shares to reflect changes in the degree of land degradation. In fact, without adequate monitoring, this instrument may instead be a disincentive to conservation.

Another possible improvement is to increase the share allocated to ecological criteria and also to include other equally ecological features of the municipalities such as sanitation and education levels. Of course, studies would need to be done to analyse the resultant effect of these changes and to verify distributional impacts and expected political barriers. Above all, it is of paramount importance to analyse the economic significance of this extra revenue in pursuing conservation activities in the entitled municipalities.

GENERAL RECOMMENDATIONS ON ECONOMIC INCENTIVES

Despite the need to correct the failures and maximize the expected successful results of the above economic incentives, a significant amount of the revenue is estimated to come from these instruments, as shown in Table 19.1 below.

One could, however, recommend a set of recommendations towards the improvement of economic incentives for forest preservation purposes in Brazil. Firstly, a solid sustainable criterion for agricultural credit incentives could be introduced. In fact, the government passed a decree in 1996 called the *Protocolo Verde* (Green Protocol) which states that any public loan or credit incentive for any kind of economic activity can only be granted if the applicant can prove that they are already complying with environmental legislation and have no outstanding environmental sanctions.[8]

Table 19.1 *Economic Incentives for Conservation in Brazil*

Instruments	Estimated Revenue (US$million)	Purpose
Fiscal compensation for conservation areas (water supply and ecosystem areas) in São Paulo, Rio de Janeiro and Paraná States.	127	to compensate municipalities for land-use restrictions for environmental purposes
Federal Reforestation Fund	7	to finance public reforestation projects
Forest tax in Minas Gerais State	12	to finance government activities on forest policy

Sources: Seroa da Motta and Reis (1994); Seroa da Motta (1996b)

In the case of industrial, commercial, and service activities, environmental operation licensing and the list of sanctioned firms controlled by environmental agencies will be the key mechanisms by which this green certification can be applied. However, for agricultural activities there is no such environmental control and farms, as previously discussed, are not obliged to comply with established technology or agroecological criteria. As already recognized in the *Protocolo Verde*, the definition of technical parameters is planned to help government banks to apply the *Protocolo*'s rules for agricultural activities.[9]

Secondly, improved controls on forest management are called for. Sustainable logging practices are, in theory, required by law. However, as already discussed, these controls are unsatisfactory because wood supplies are readily obtainable through agricultural expansion. Even with the introduction of sustainable criteria for agricultural practices, land clearing will continue to be a major source of wood supply. Removing clearing criteria for titling and taxation is also regarded as an incentive for land concentration.[10]

Therefore, a promising policy alternative for forestry is a system of public concessions similar to those applied in countries such as Canada and Sweden. Concessions are long-term leasing contracts for large tracts of forest, allocated by international auction, to private corporations with clauses specifying accepted conditions on the use of land and natural resources.[11] Non-compliance with the sustainable practices defined in concession licensing would be subject to sanctions and termination of the concession. Supervision and monitoring of these concessions could be shared with NGOs. This scheme is particularly feasible, for example, in Amazonia, where there are still large unclaimed areas – more than 25 per cent according to recent surveys of the Agrarian Reform Institute (Incra).

In addition to some technical features which need to be addressed (including managerial practices, concession periods, stumpage fees, and so on), such changes in property right assignments might encounter political barriers over issues such as land concentration, international ownership and agricultural activity restrictions.

Thirdly, a fiscal instrument such as the forestry tax, previously described for the case of Minas Gerais, could be of immense importance in regulating current forestry activity. This instrument would need to be applied nation-wide and be defined according to marginal user cost curves, to encourage a desirable level of forestry activity or conservation. Since such economic instruments are difficult to enforce, due to institutional fragility, the revenue collected must be partially earmarked to environmental agencies and research centres where forest research is undertaken.

Fourthly, fiscal compensation from VAT revenue, already in place in some Brazilian states, could be nationally applied with compensation levels set on the basis of output forgone as well as on area and enforcement criteria.

Fifthly, biodiversity research could be encouraged by offering incentives for joint ventures between Brazil and international companies to undertake such work. Restricting such incentives to only Brazilian firms would limit their impact. A similar situation can be seen in the very generous credit and fiscal incentives, which have been offered for technology R&D. In encouraging joint ventures, the issues of sharing property rights and compensation would need to be addressed.

Finally, it should be recognized that increasing public awareness of biodiversity conservation issues is vital to politically sustain current and future conservation policies in Brazil.

NOTES

1   Instituto de Pesquisa Economica Aplicada.

2   See Seroa da Motta (1993).

3   See Seroa da Motta (1996a) for detailed indicators on deforestation in Brazil and their analysis.

4   See Bonelli and Ramos (1993).

5   According to the Brazilian Forest Code, clearing in the Amazonian region for agricultural purposes was allowed on up to 50 per cent of the property area (see Seroa da Motta, 1993). New legislation was passed in August 1996, reducing the clearing allowance to no more than 20 per cent of the property area, subject to zoning indicators and the current size of the already-degraded area.

6   This norm was created by the former Brazilian Institute for Forestry Development which is today part of IBAMA.

7   As in the case of the national reforestation tax, data recovery from official reports is very difficult, due to the inflation rates prevailing in Brazil until 1995.

8   In fact, the existing legislation already includes this restriction, although it did not mention sanctions and pre-operation licensing was accepted as environmental compliance. Therefore, *Protocolo Verde* can be understood as a regulation of this legislation.

9   Economic-ecological zoning can be applied in this case.

10   Private forest reserves can be exempted from property tax if owners comply with very strict forestry standards. However, since land tax levels are relatively low (representing less than 0.01 per cent of GDP) and not fully enforced, such incentives are of very limited use.

11   See Seroa da Motta (1994).

REFERENCES

Bonelli, R and Ramos, L (1993) Distribuição de Renda no Brasil: Avaliação das Tendências de Longo Prazo e Mudanças de Desigualdade Desde Meados dos anos 70. Texto para Discussão 288, IPEA/DIPES, Rio de Janeiro.

Seroa da Motta, R (1993) Past and Current Policy Issues Concerning Tropical Deforestation in Brazil, Kiel Working Paper 566, The Kiel Institute of World Economics, Kiel.

Seroa da Motta, R (1994) Política e gestão florestal; in: O Brasil no Fim do Século: Desafios e Propostas para a Ação Governamental. IPEA/DIPES, Rio de Janeiro.

Seroa da Motta, R and Reis, E J (1994) The Application of Economic Instruments in Environmental Policy: the Brazilian Case. Paper presented at the OECD/UNEP Workshop on The Use of Economic Policy Instruments for Environmental Management, mimeo, Paris, 26–27 May, 1994.

Seroa da Motta, R (1996a) Indicadores Ambientais no Brasil: Aspectos Ecológicos, de Eficiência e Distributivos, Texto para Discussão 403, IPEA/DIPES, Rio de Janeiro.

Seroa da Motta, R (1996b) The Economics of Biodiversity in Brazil. Paper presented at the OECD International Conference on Biodiversity Incentive Measures, Cairns, 25–28 March 1996.

GLOSSARY

air dispersion model: a computer program that is used to estimate air quality concentrations from pollutants that would be emitted from new (or existing) emissions sources. Examples of emissions sources include stack emissions from industry and areawide emissions from groups of sources (such as automobiles or woodstoves).

ambient air quality standards: standards set in pollution control regulations, which establish the highest allowable concentration of specific pollutants in the ambient air.

assimilative capacity: the capability of the environment to take wastes and convert them back into harmless or ecologically useful products (Pearce and Turner, 1990).

Biochemical Oxygen Demand (BOD): an index of water pollution, which represents the content of biochemically degradable substances in a sample of water. A test sample is stored in darkness for five days (unless another number of days is specified); the amount of oxygen taken up by the microorganisms present is measured in grams per cubic metre. This test, while no longer considered an adequate criterion by itself for judging the presence or absence of organic pollution, is still widely used, particularly in sewage treatment (Gilpin, 1996).

carbon sequestration: the long-term storage of carbon in soils, vegetation, wetlands and oceans, which can help limit the global warming effects of anthropogenic release of carbon dioxide into the atmosphere.

carrying capacity: the rate of resource consumption or waste discharge that can be sustained indefinitely in a defined impact region without progressively impairing bioproductivity and ecological integrity (UNEP, 1996).

catalytic converter: a device attached to internal-combustion engines to reduce emitted nitrogen oxides, carbon monoxide and hydrocarbons in the exhaust. They still allow carbon dioxide emissions (Gilpin, 1996).

Chemical Oxygen Demand (COD): the weight of oxygen taken up by the total amount of organic matter in a sample of water. This measure is used to assess the strength of sewage and wastewater. The biochemical oxygen demand and the COD tests are of equal importance in wastewater treatment (Gilpin, 1996).

chi-square tests: a measurement to indicate the likelihood that the results of a survey could have occurred by pure chance. Social scientists look for results in which the possibility is less than 5 per cent that the results can be explained by chance. See *confidence level*. In econometric regression analysis, the chi-square *goodness of fit* test is one of several tests applied to check the normality of the disturbance term (Gujarati, 1995).

CO: carbon monoxide.

CO_2: carbon dioxide.

command-and-control approach: an approach to environmental protection and pollution control which uses regulations to restrict polluting activities. Command-and-control policies are based on fixed pollution standards that must be met by all polluters independent of relative abatement and damage costs. An alternative approach to pollution control uses *economic instruments*.

confidence level: in statistical analysis, the desired probability, selected by the investigator, that the universe mean will be included in the calculated limits. For most ecological work, 95 per cent (0.05) or 99 per cent (0.01) confidence limits are considered satisfactory.

earmarking: used here to refer to the common practice of directing the revenue raised from the collection of environmental charges to specific purposes and projects (often through environmental funds). Economists have traditionally not favoured earmarking because it introduces distortions into the economic system, that is, it forces revenues to be used for specific purposes rather than be allocated to their highest social use. Earmarking is also a potential source of inefficiency in that it prevents the maximization of total returns, by preventing equalization of marginal returns across uses. Modern social choice analysts argue that earmarking fulfills a democratic requirement since it allows voters to keep track of how revenues are being used (Pearce and Warford, 1993).

economic efficiency: an economic efficient level of pollution is that level of pollution at which the marginal net private benefits of the polluting firm are just equal to the marginal external damage costs (Pearce and Turner, 1990).

economic instruments: policy measures which make use of market forces to encourage polluters and users of environmental resources to adopt more environmentally-friendly behaviour. Economic instruments, or *market based instruments* as they are sometimes called, include *user charges*, *product charges*, and *tradeable emission permits*. An alternative approach to pollution control is regulation-based, and is known as a *command-and-control* approach.

effluent charge: a charge levied on polluters discharging into water bodies, based on the quantity and/or quality of the effluent. Effluent charges generally have a revenue-raising function, and the revenue raised is usually *earmarked* for specific environmental purposes.

emission charge: as for *effluent charge*, in this case applied to air pollution. Similarly to effluent charges, emission charges generally have a revenue-raising function as well as a pollution control function.

emission reduction credit: a credit made available, as part of a *pollution offset system*, to a polluting source that has reduced its emissions more than is required by the standard set. A polluter may also earn an emission reduction credit by undertaking environmentally beneficial activities, even in countries other than the one in which they are a pollution source. In some cases, emission reduction credits can be traded, as *tradeable emission permits*.

EPA: Environmental Protection Agency. These government agencies responsible for environmental protection are increasingly common in developing countries. EPAs have generally focused on *command-and-control* measures, though more and more are now turning to *market- based instruments*.

excise tax: a tax imposed on the importation or sale of specific commodities. Traditionally, excise taxes or duties have related to domestically produced goods, and customs duties to imported goods, but the distinction is not rigid. The motives for imposing these charges are generally: (i) to raise revenue; (ii) to protect domestic industries; and (iii) to give preference to imports from certain countries (Pearce, 1992).

externality: an unintended (negative or positive) impact on a third party's welfare, that is brought about by the action of an individual or firm, and is neither compensated nor appropriated (Pearce and Warford, 1993).

external cost: an external cost arises when an activity by one agent causes a loss of welfare to another agent, and this loss is uncompensated (Pearce and Turner, 1990).

free-rider problem: a general problem in the management of public goods, whereby some individuals avoid paying for their use of the good in question.

goodness of fit: the magnitude of R^2, which is the coefficient of determination (supposed to represent the proportion of the variation in the dependent variable 'explained' by variation in the independent variable).

Gross Domestic Product (GDP): a measure of the total production and consumption of goods and services in a country.

groundwater: the supply of fresh water under the earth's surface. Groundwater is stored in porous layers of underground rock called aquifers. See *surface water* (Raven et al, 1993).

internalization: a situation where an *externality* is taken into account and output of the offending good (eg pollution) reduced to its optimal level, with the optimal amount of the externality still in existence; ie where the cost of reducing the externality by a further unit exceeds the benefit from so doing. For example, introducing emission or effluent charges is one means of internalizing the external costs of pollution.

least cost: *economic instruments*, if properly designed, can possess the *least cost* property, ie they can achieve the predetermined environmental standards at a minimum social cost. This is the case for *pigouvian taxes* (Baumol and Oates, 1988).

marginal social cost: the extra *social cost* of producing an extra unit of output (including, for example, pollution).

marginal value: the extra value (benefit) obtained from an extra unit of any good.

market-based instruments: see economic instruments.

NGOs: non-governmental organizations.

NO$_x$: nitrous oxides, ie NO (nitrogen monoxide) and NO$_2$ (nitrogen dioxide).

O$_3$: ozone, a blue gas which is a human-made pollutant in one part of the atmosphere (the troposhere) but a natural and essential component in another (the stratosphere).

OECD: the Organization for Economic Co-operation and Development. OECD's fundamental task is to enable members to consult and cooperate with each other to achieve the highest sustainable economic growth in their countries, and improve the economic and social wellbeing of their populations. The OECD offers advice and makes recommendations to its members to help them define their policies. On occasion it also arbitrates negotiations of multilateral agreements and establishes legal codes in certain areas of activity. OECD currently has 29 member countries, including western European countries, North America, Japan, Australia, New Zealand, Finland, Mexico, the Czech Republic, Hungary, Poland and Korea (OECD, 1998).

pareto optimal: a pareto optimal situation is one in which it is impossible to make any individual better off without making someone else worse off (Pearce and Turner, 1990).

pH: a number from 0 to 14 that indicates the degree of acidity or alkalinity of a substance.

pigouvian concept/tax/level: a tax on the polluter based on the monetary value of damage done (ie based on the external costs of their polluting behaviour). This theoretically ideal Pigouvian tax achieves an optimal level of externality where the marginal costs of pollution abatement are equated to the marginal pollution damage costs. Pigouvian taxes (if they could be determined in practice) would, it is presumed, lead to *Pareto optimal* levels of the activities (Markandya and Richardson, 1992).

Polluter Pays Principle (PPP): a principle formulated by the OECD in the early 1970s, stating that 'the polluter should bear the expenses of carrying out the (pollution prevention and control) measures decided by public authorities to ensure that the environment is in an acceptable state' (OECD, 1972). This statement does not make it clear whether the polluter should also be responsible for the pollution damage that their emissions still cause when the environment has reached an 'acceptable state'. To clarify this situation, a later statement was issued by the OECD which reads 'if a country decides that, above and beyond the costs of controlling pollution, the polluters should compensate the polluted for damage which would result from residual pollution, this measure is not contrary to the PPP, but the PPP does not make this additional measure obligatory' (OECD, 1975). In practice, no country has attempted to implement the extended polluter pay principle due to fierce resistance from industry to any proposals for pure incentive charging on the full amount of the effluent load (Markandya and Richardson, 1992).

pollution offsetting: under a pollution offset system, permits are issued to polluters, based on the emission standards in operation and the desired amount of emission reduction. Major new or expanding sources in areas where the air quality if worse than the ambient standards are then required to secure sufficient offsetting emission reductions (by acquiring *emission reduction credits*) from existing firms, so that the air is cleaner after their entry of expansion than before (Tietenberg, 1992).

ppm: parts per million, the number of parts of a particular substance found in one million parts of air, water, or some other material.

price differentials: differences in the prices of (closely-related) products. For example, a price differential could be created for leaded and unleaded gasoline, to make unleaded gasoline significantly cheaper than leaded gasoline, to encourage vehicle owners to switch to unleaded fuel.

product charge: an addition to the price of a product which is polluting (for example, gasoline) or difficult to dispose of (for example, packaging). Product charges, as with most types of environmental charge, generally have a revenue-raising function as well as an environmental protection function.

savannah: a tropical grassland with widely scattered trees or clumps of trees; found in areas of low rainfall or seasonal rainfall with prolonged dry periods.

self-reporting: reporting by polluting firms themselves of their level of emission or discharge. As the quantity of pollution reported determines the amount of charge levied, self-reporting often creates problems of under-reporting, thus reducing the amount of charge revenue collected.

sewage: the contents of sewers carrying the waterborne wastes of a community.

sewerage: a physical arrangement of pipes and plant for the collection, removal, treatment, and disposal of liquid wastes (Gilpin, 1996).

social benefit: the sum of the gains or benefits deriving from an activity or project to whomsoever they accrue. Occasionally used to describe 'external benefits', ie benefits to others, rather than to the individual who predominantly enjoys the private benefit (Pearce, 1992).

social cost: the social cost of a given output is defined as the sum of money which is just adequate when paid as compensation to restore to their original utility levels all who lose as a result of the production of the output. The social cost is the *opportunity cost* to society, rather than to just one firm or individual. One of the major reasons why social costs differ from the observed private costs is due to the existence of *externalities* (Pearce, 1992).

solid waste: waste comprised of *municipal solid waste* (generated primarily in homes and businesses) and *non-municipal solid waste* (generated by industry, agriculture and mining).

surface water: fresh water found on the earth's surface in streams and rivers, lakes, ponds, reservoirs and wetlands. See *groundwater* (Raven et al, 1993).

Total Suspended Solids (TSS): the total solids in a liquid that can be removed through sedimentation or filtration; a measure of TSS is used to help assess water quality.

tradeable emission permits: permits issued to polluting enterprises ('permitting' them to pollute up to a certain level), which are allowed to be bought and sold. The rationale for establishing an emissions trading system is that polluters with high abatement costs will prefer to buy the permits, while low abatement cost polluters will sell permits. See *pollution offset system*.

tragedy of the commons: a tragedy of overgrazing and lack of care which resulted in erosion and falling productivity of the English commons, prior to the enclosure movement, when grazing rights became restricted to the few. As the number of animals became more than the village commons could support, no one had any interest in ensuring the future productivity of this resource. Today, on a much larger scale, the natural resources of air and water may be regarded as the 'commons of the world'. In 1968, the theme was elaborated by Garret Hardin of the University of California and the term is associated with his name (Gilpin, 1996).

transboundary pollution: pollution arising from one country and causing damage in another country.

user charge: an environmental charge levied on the users of natural resources, most commonly water. Another form of user charge is an entrance fee for visitors ('users') of recreational facilities, such as national parks.

volatile organic compounds (VOC): any organic compounds that participate in atmospheric photochemical reactions.

REFERENCES

Baumol, W J and Oates, W E (1988) *The Theory of Environmental Policy.* Second edition, Cambridge University Press, Cambridge, U.K.

Gilpin, A (1996) *Dictionary of Environment and Sustainable Development.* John Wiley and Sons, Chicester.

Gujarati, D N (1995) *Basic Econometrics*. Third Edition. McGraw-Hill International Editions, Economic Series.

Markandya, A and Richardson, J (1992) *The Earthscan Reader in Environmental Economics*. Earthscan Publications, London.

OECD (1998) Webpage 'About OECD', edited on 31 July, 1998.

OECD (1972) Annex to the Recommendation on Guiding Principles Concerning International Economic Aspects of Environmental Policies. OECD, Paris.

OECD (1975) *The Polluter Pays Principle: Definition, Analysis, Implementation.* OECD, Paris.

Pearce, D W (ed) (1992) *Macmillan Dictionary of Modern Economics*. Fourth edition, Macmillan Press, London.

Pearce, D and Turner, R K (1990) *Economics of Natural Resources and the Environment*. Johns Hopkins Press, Maryland.

Pearce, D W and Warford, J J (1993) *World Without End: Economics, Environment, and Sustainable Development*. Published for The World Bank by Oxford University Press, New York.

Raven, P H, Berg, L R and Johnson, G B (1993) *Environment.* Saunders College Publishing, Harcourt Brace Jovanovich College Publishers, Orlando.

Tietenberg, T H (1992) Economic Instruments for Environmental Regulation; in: Markandya and Richardson, op cit.

UNEP (1996) *Environmental Impact Assessment: Issues, Trends and Practice.* Prepared for the Environment and Economics Unit of UNEP, by Scott Wilson Resource Consultants. UNEP, Nairobi.

INDEX